Precultural Primate Behavior

Symposia of the Fourth International Congress of Primatology

Portland, Ore., August 15–18, 1972

Vol. 1

Main Editor: WILLIAM MONTAGNA
Oregon Regional Primate Research Center, Beaverton, Ore.

S. Karger · Basel · München · Paris · London · New York · Sydney

Precultural Primate Behavior

Volume Editor: EMIL W. MENZEL, jr.
State University of New York at Stony Brook, N.Y.

With 39 figures and 14 tables

S. Karger · Basel · München · Paris · London · New York · Sydney 1973

Symposia of the Fourth International Congress of Primatology

Vol. 1: Precultural Primate Behavior.
   XVI + 252 p., 39 fig., 14 tab., 1973. 3–8055–1494–8
Vol. 2: Primate Reproductive Behavior.
   X + 125 p., 30 fig., 29 tab., 1973. 3–8055–1495–6
Vol. 3: Craniofacial Biology of Primates.
   X + 268 p., 66 fig., 29 tab., 1973. 3–8055–1496–4
Vol. 4: Nonhuman Primates and Human Diseases.
   XII + 149 p., 27 fig., 19 tab., 1973. 3–8055–1497–2

S. Karger · Basel · München · Paris · London · New York · Sydney
Arnold-Böcklin-Strasse 25, CH-4011 Basel (Switzerland)

© Copyright 1973 by S. Karger AG, Verlag für Medizin und Naturwissenschaften, Basel
Printed in Switzerland by Coop-Schweiz, Basel
Blocks: Steiner & Co., Basel
ISBN 3–8055–1494–8

Contents

Foreword to Symposia Proceedings, Volumes 1–4

Although primatology has a long and exciting history, it has come into bloom only during the last few decades. Perhaps the reason is that we have begun to appreciate the importance of using nonhuman primates in many biological and biomedical disciplines. In recent years we have seen a steady increase in the number of animals used and in publications and conferences on primates. The trend has been slow, and its maturation is welcome; but it poses a number of very serious problems. One of these, perhaps the most important, is the lamentable and promiscuous use of these animals without a clear knowledge or appreciation of their biological value, hence of the care they require; this alone has seriously threatened the survival of many species. In the past (and, unfortunately, too often in the present), biomedical scientists have requested and used large numbers of them with the same casualness as they would laboratory rats. Part of this disregard for the value of these animals is the attitude that experimental observations on primates are somehow more prestigious or meaningful than those on other animals, even when such studies barely meet the standards required of the biological sciences. The increasing demand for experimental nonhuman primates engendered by such attitudes has placed them all on the growing list of doomed species. Since primatologists are almost the only scientists to be aware of this danger, the real responsibility for safeguarding dwindling populations of feral primates rests squarely on them. It is they who must generate concerted action at Congresses and must assume the moral responsibility and direction for preserving wild and captive populations of animals and for insisting that they be used as special beings.

Another unhappy offshoot of growth and maturation is the size and quality of national and international meetings of primatologists. With the proliferation of more and more primatological studies and the continuing accumulation of information, the traditional organizational format is no longer effective. Basically unselective and too crowded for any social interaction or meaningful interchange of scientific information, that type of format has had its day; and like all else that is outmoded and useless had best be avoided in the future. Aware of the need to revamp the overall plan of the Congress, the Chairman and Committee of the Fourth International Congress of Primatology (IPC), held in Portland, Oreg., in August 1972 had the task of facing up to the major issue of communication and of focusing on the important and excluding the peripheral. Thus, the format they chose, which greatly reduced the number of papers presented and permitted only two, or in a few cases, three concurrent sessions, was a departure from that of the three previous Congresses of the International Primatological Society (IPS).

At the core of the program were four day-long symposia of invited contributions in Precultural Primate Behavior, Craniofacial Biology of Primates, Primate Reproductive Behavior, and Primate Medicine and Pathology, organized and chaired, respectively, by Drs. EMIL W. MENZEL, jr., M.R. ZINGESER, C.H. PHOENIX, and W.P. McNULTY, jr. Most of the symposia presentations are published in this series of four volumes, edited in each case by the respective chairman. In each of these books, which identify and channel four major areas of primatological investigation, the contributing authors review the various facets of their topics and add new information. Organizers of future Congresses of the IPS may want to follow this format, choosing other subjects as major themes: for example, Human Evolution, Molecular Biology, Neurophysiology. An alternation of principal themes every two years should help to bring the major areas of research into clear focus.

The Fourth IPC and the Congress *Proceedings*, which include the four volumes of this series and the March 1973 issue of the *American Journal of Physical Anthropology*, are dedicated to SHERWOOD L. WASHBURN, Professor of Anthropology at the University of California at Berkeley, in recognition of the great impact he has made on modern approaches to anthropology and to primatology in particular. WASHBURN has brought such broad ranges of biological disciplines as anatomy, physiology, genetics, demography, sociology, behavior, culture, and human biology in general to bear on the single problem of human evolution. His bibliography gives some insight into the evolution of the man himself. In his first publications, for example, he was preoccupied with functional anatomy. Later, extensive field trips to East and South Africa, Ceylon, Thailand, Borneo and other places broadened his perspective; and he wrote numerous reports on social structure and behavior and communication in nonhuman primates. As his perception of nonhuman primate society, behavior, and communication deepened, he became directly engrossed in human evolution, which he contends is inseparably bound up with the evolution of culture.

WASHBURN has not only opened up new vistas of human evolution but has trained a large number of disciples, many of whom have achieved eminence even at an early age and have been inspired and urged by him to explore fresh approaches of their own. These new vistas and fresh approaches are the key words in the organization of the IVth IPC and in these four volumes of the Congress Symposia Proceedings.

The generous financial support of the Congress by the Psychobiology Program of the National Science Foundation, The Wenner-Gren Foundation for Anthropological Research, The Medical Research Foundation of Oregon, the International Union of Biological Sciences, and the International Primatological Society is most gratefully acknowledged. I wish also to thank the members of the Congress Organizing and Program Committees, Drs. W.H. FAHRENBACH, THEODORE GRAND, WILBUR MCNULTY, EMIL MENZEL, CHARLES PHOENIX, ADRIENNE ZIHLMAN, and M.R. ZINGESER, and JOEL ITO, CLAIRE LYON, and ISABEL MCDONALD; my secretary, MAXINE ALLEN; the Congress Coordinator, ROSEMARY LOW; and all the other colleagues at the Oregon Regional Primate Research Center too numerous to mention, who helped the Congress attain a distinction that can only arise from an ideal of excellence and a superlative display of teamwork.

October 1972   WILLIAM MONTAGNA
                                      Director, Oregon Regional Primate
                                      Research Center, and
                                      Chairman, IVth International Congress
                                      of Primatology

Preface

On August 15, 1972, at the Fourth International Congress of Primatology in Portland, Oreg., a full-day Symposium on primate behavior and proto-culture was held. I served as Chairman of the Symposium and HANS KUMMER as Co-Chairman, but its real instigator was HELMUT HOFER. I sincerely thank Dr. HOFER for the initial idea and his subsequent advice in the selection of topics and speakers. Most of the papers in this volume are based directly on those read at the Symposium, but we were fortunate enough to entice Drs. ITANI and NISHIMURA and Dr. ROGERS to write additional articles especially for inclusion in this volume; and I was also asked to add one of my own. Although many gaps in coverage still remain, the volume exceeds all our prior expectations; and the contributions also probably render these introductory remarks obsolete.

For the comparative analysis of 'culture' perhaps the principle should be the more species the merrier; but in view of the limited number of articles that could be accommodated here and the dearth of information on most species of primates, it seemed wise to concentrate on two of the best-studied species, namely Japanese monkeys and chimpanzees. This bias is to some extent counteracted by one of the single-species papers (KUMMER'S) and by those which give topic-oriented rather than species-oriented surveys (those of COUNT, GARTLAN, EISENBERG, and RENSCH). However, no account of modern studies on protoculture in animals would be complete without consideration of the pioneer Japanese work on *Macaca fuscata;* and the paper by Drs. ITANI and NISHIMURA not only summarizes the empirical findings of this research, but also gives the reader many insights into the people who did it and provides a fascinating history of their scientific debates. To Dr. KINJI IMANISHI, guiding spirit of research, and Dr. ITANI, especial thanks!

As Dr. HOFER and I saw it in 1970, when we first proposed the Symposium, its task would be to look at 'culture-like' phenomena in primates, including man, closely enough to be aware of the important differences between species, but at the same time to take a broad enough view to allow one to see similarities and continuities wherever they appear. In short, the Symposium would place the concept of culture into a primatological rather than a strictly anthropological perspective. It should not be averse even to taking current primatology to task for failing to perceive the still broader perspective of mammalogy. When one rises to the level of generality at which one can discuss all mammalian societies within the same frame of reference, it might of course be that the term 'culture' is appropriate only in a very loose adjectival sense; therefore, a term such as protoculture, preculture, infraculture, or 'biological bases of culture' can be substituted if one so desires. Whether or not any of these terms can be defined in a fashion that is acceptable to all is another problem. This problem might disturb deductively-oriented theorists; but it should not disturb the inductively-oriented researcher who wants to know first of all, what do animals and men *do*, and why? Dr. COUNT's paper in particular should be extremely useful in pointing out leads for the inductively-oriented investigator to follow, but all of the papers in this Symposium will be suggestive.

If the present Symposium creates some strain on an anthropocentric definition of culture, we might take heart from EINSTEIN and INFELD [1951, pp. 158–159]. To paraphrase them: Creating a new primatological theory is not like destroying an old barn and erecting a skyscraper in its place. It is rather like climbing a mountain and gaining new and wider views and discovering unexpected connections between our starting point and its surrounding environment. The path from which we started still exists and can be seen although it appears smaller and forms a small part of our broad view gained by the mastery of the obstacles on our adventurous way up. A primatological theory would show the merits as well as the limitations of anthropological theories and allow us to regain our old concepts from a higher, more fundamental level.

Dr. SHERWOOD WASHBURN [1973], in his address to the Fourth International Congress of Primatology, expressed a rather different opinion about the nature of theories when he maintained that a theory of man or of culture is not 'out there', like a geographical landmark; rather, it is a pure creation of human theorists. This formulation may be true; but, I would counter, it is not exactly the point. Animals and men are palpably 'out there', and their behaviors are tangible events that occur in particular places at particular times.

It is these *phenomena* which interest us and which we seek to understand. Phenomena and not theories are the measure of things real; and if it is possible to account for them directly, so much the better. In this respect, a certain degree of 'naive realism' is not only warranted but essential for a practicing science of culture [see, e.g., HALL, 1959].

The 'old barn' from which our metaphorical tour starts out will be indicated in passing by our tour guides, especially the first one, Dr. COUNT, an anthropologist, and the last one, Dr. EISENBERG, a zoologist. They were asked specifically to consider the relative uniqueness of man (COUNT) and of primates (EISENBERG) in a general account of culture-like phenomena. Other guides will tell us about places and scenes along the trail that are hopefully on the same mountain or chain of mountains. My role in this metaphor was that of the travel agent; and if I have omitted some of the places that you feel should have been included, I apologize. I do not personally know much of the terrain and most of my knowledge of it has been gained in the armchair. Furthermore, if anything, I prefer watching and reading about the animals and the landscape to reading about people; and animal behavior would be of interest to me even if it shed no light at all on what was going on behind the barn.

Since most of the authors are students of animal behavior, they perforce shy away from limiting their treatment of culture to its dictionary definition of 'traditional, that is, historically derived ideas and their attached values'. As SKINNER [1971] would say, even those who talk about human culture do not worry greatly if they occasionally stray from the realm of the mentalistic and uniquely humanoid. They tell us about how people live, how they handle and discipline their children, how they secure food and other basic necessities, what kinds of dwellings (if any) they live in, what sorts of clothes (if any) they wear, what games they play, how they regulate their disputes, how they cooperate against common enemies, what rules govern their priority of access to lands and resources, and so on. These are the customs, the customary behaviors, of a society. More often than not the student of culture cannot tell us precisely how these customary behaviors originated, and therefore we may properly be skeptical about any attempt to limit the concept of culture to symbolically-transmitted events alone [see also HALL, 1959]. To specify the nature and ontogenesis of any given behavior precisely would require experimental analyses of the contingencies that generated that behavior; and we have scarcely begun such analyses.

In general, the 'complete' description and analysis of culture-like phenomena (whether in animals or in man) has three sub-goals that may be ordered in step-wise or hierarchical fashion:

I. The first subgoal is to describe what behaviors are usual, typical, or customary in a given society. This specifies only what the members of the society do, not (necessarily) how they came to do it. Most of the empirical papers in this symposium are useful principally at this level. For example, Dr. VAN LAWICK-GOODALL tells us how wild chimpanzees learn from each other; Dr. KUMMER tells us how baboon males govern themselves; Dr. RENSCH tells us what games animals play; Drs. NISHIMURA and STEPHENSON and I tell how monkeys and apes communicate with each other. Since a majority of culture-like phenomena occur in a social context and involve communication of some sort, the heavy emphasis on social behavior and communication is deliberate.

A word must be said about the rather slippery term 'behavior'. If anyone argues that most behavior (in monkeys or man) is innate, you can be almost sure that he is talking about 'behavior' of a relatively molecular sort (e. g., the ethologists' fixed motor patterns); and if anyone argues that most behavior is learned or culturally-acquired, you can be almost certain that he is talking about molar performances which cannot be defined without taking into account their consequences, their contexts, or the objects toward which they are oriented. I believe that this point is made, either explicitly or implicitly, by nearly all of the authors. To some researchers this might suggest that cross-species comparisons must be based on comparable levels of analysis and similar 'sizes' of behavioral units. But the problem is not so simple. It is more than likely that the 'size', as well as the 'shape', of the so-called 'natural units of behavior' [ALTMANN, 1965; EISENBERG, this volume] varies markedly from man to chimpanzee to marmoset, or even from human infant to human adult [see BARKER and WRIGHT, 1955]; thus, to use comparably-sized behavior units in all cases as a basis for talking about culture is tantamount to throwing out a crucial aspect of the problem of how behavior is organized. It is not the motor elements of individual behavior but the organization of whole societies for which the student of culture must ultimately account. What are clearly needed, but are as yet unobtainable in primatology, are adequate objective methods for dealing with all levels of organization *simultaneously* [MILLER *et al.*, 1960; MENZEL, in press].

II. The second task is to identify which 'customary behaviors' originate through individuals' social experience in a particular society, i. e., behaviors that can be called 'traditional' rather than innate or learned solely through individual acquisition of information. There are four major ways of proceeding:

(a) The deprivation experiment [LORENZ, 1965; ROGERS, this volume].

(b) The cross-fostering or home-rearing experiment [HEWES, this volume].

(c) 'Cross-cultural comparisons' and examination of the same species under varied natural conditions [GARTLAN, this volume; ITANI and NISHI-MURA, this volume].

(d) Studies of how information exchange or social learning takes place in 'normal' individuals of a given society [STEPHENSON, NISHIMURA and VAN LAWICK-GOODALL, this volume].

III. The final task is to draw together all of these sources of information regarding a given society and to specify how the various behaviors are inter-related, the conditions under which each appears or fails to appear, and the mechanisms of information exchange that are necessary and sufficient for the behaviors to pass from one generation to the next.

If this stage of analysis can ever be reached, the relationships between genetic, developmental, learned, and specifically socially-learned aspects of culture will fall into place quite naturally. For the goal of any anthropologi-cally-oriented or primatologically-oriented or biologically-oriented science is only one facet of the general problem of evolution: to understand how animals and people came to be as they are and behave as they do in the world at large. Needless to say, no single symposium or book will take us very far toward our common goal. But the amount of information in a volume is best assessed by the degree to which it reduces one's prior ignorance; and by that criterion, at least for me these papers have all been highly informative.

The symposium on which this volume was based would not have been possible without travel funds for many of the authors from the National Science Foundation and the Wenner-Gren Foundation. Editing costs and travel funds for the chairman were further supported in part by the National Science Foundation grant GU 3850 to the Psychobiology Program of the State University of New York at Stony Brook. Special thanks are due to ROSEMARY LOW, RUTH CONNELLY, PAT CARL, and HARRIET ANNE MENZEL for their help with the editing and typing.

References

ALTMANN, S. A.: Sociobiology of rhesus monkeys. II. Stochastics of social communication. J. theor. Biol. 8: 490–522 (1965).

BARKER, R. G. and WRIGHT, H. F.: Midwest and its children (Harper & Row, New York 1955).

EINSTEIN, A. and INFELD, L.: The evolution of physics (Simon & Schuster, New York 1951).

HALL, E. T.: The silent language (Doubleday, Garden City 1959).

LORENZ, K. Z.: The evolution and modification of behavior (Chicago University Press, Chicago 1965).

MENZEL, E.W.: A group of young chimpanzees; in SCHRIER and STOLLNITZ Behavior of nonhuman primates, vol. 5 (Academic Press, New York, in press).

MILLER, G.A.; GALANTER, E., and PRIBRAM, K.H.: Plans and the structure of behavior (Holt, Rinehart & Winston, New York 1960).

SKINNER, B.F.: Beyond freedom and dignity (Knopf, New York 1971).

WASHBURN, S.L.: The promise of primatology. Proc. 4th Int. Congr. Primat., Portland, Oreg. 1972. Amer. J. phys. Anthropol., (in press, 1973).

Stony Brook, New York   EMIL W. MENZEL, jr.
October, 1972

Symp. IVth Int. Congr. Primat., vol. 1: Precultural Primate Behavior,
pp. 1–25 (Karger, Basel 1973)

On the Idea of Protoculture

E. W. Count

Professor *Emeritus*, Hamilton College, Clinton, New York

Mr. Chairman, Fellow-Primates: What were more welcome to a student
of his own genus than to have the privilege of appealing to colleagues whose
study is the order within which his genus belongs for help in placing Man, not
in Nature – that were too ambitious – but within this very restricted domain of
it, in ways yet to be defined! To my friend, Dr. MENZEL, my thanks for the
opportunity; thanks further, to the goodly company of yeomen who have
wrought to make the opportunity a practicable thing; particularly to that
gracious one among them, ROSEMARY LOW, whose labors shall not yet be
ended when this presentation has ended. And in a very special way, now, to
Mrs. LITA OSMUNDSEN, Director of Research of the Wenner-Gren Foundation.
For her long and personal interest in the research itself, from which this paper
has been abstracted, I beg to dedicate it to her, both as it is read and when it
has entered print.

When Dr. MENZEL first invited it, I pleaded but a very interested on-
looker's knowledge of primatology. He responded to the effect that he considered my ignorance an asset. Dr. MENZEL is indeed a mighty persuader.
Suddenly I found myself a companion to Socrates. How could I resist the
temptation to flourish my asset before you? On, then, with the *antipasto;* let
the *pièce de résistance* follow, as the train of symposiants takes over.

This preliminary appeal, however, is a challenge rather than a petition;
for it is diametrically the contrary to what you might expect of an anthropologist. Let me state the conclusion at which I hope we shall arrive: it is the
primatologists who should first give us a working definition of 'protoculture'
from which the anthropologists should seek to work out a defintion of 'culture.
This decidedly is not the way these things are usually done. But 'culture' first,
'protoculture' therefrom, is anthropocentric. 'Protoculture' thereby becomes

identified via a method of deficit and beats against the current of evolutionary process. To be sure, a definition of culture is a sound first heuristic step; nevertheless, it eventually becomes as awkward for tracing evolutionary processes as driving a car in reverse while watching the driver's mirror.

1. Toward Clarification of a Problem:
What is 'Culture'?

1.0 Several years ago, over preprandial cocktails, I was trying to expound to a colleague, a justly celebrated ethnographer, the reasons why we anthropologists must become better than casually acquainted with the life-modes of nonhuman primates. (Some of my friends here will guess, correctly, that I was sawing on my pet theme of the 'biogram'.) She had never pretended to any sophistication about matters biological; still, I thought things were going very well, until she interrupted with the question, 'Do apes have culture?' My generalship has none of the headlong quality of a Patton or a Jeb Stuart; I withdrew from the field, I hope in good order; we went to dinner and talked of other things.

1.1. On reflection, it appears that both of us were victims of historical circumstances, our impasse rooted in the fact that cultural anthropology has no helpful definition of culture. It has contributed this major concept to modern thought, a great achievement indeed; but you might as well ask a theologian or a social science specialist for a definition of religion. Should you do so, you would find either one better at talking to himself. Two decades ago, Kroeber and Kluckhohn reviewed critically 164 to 300 definitions of culture since 1871. (The numerical discrepancy already bespeaks the quandary.) At the times of their deaths, they were still collecting.

One can accept all these definitions, more or less; yet still they offer no help for our theme of the day. For they all premise the uniqueness of man, and then they seek to pinpoint what that is. Within the requirements of a cultural anthropology, they may have a *heurisma*. But they shut the gate on investigations into any and all processes that must have brought the human life-mode out of the infra-human. Their philosophical principle is pre-Darwinian. This is basically the reason, I suppose, that when cultural anthropology adopts the term 'evolution', it changes the meaning to apply to the cultural elaborations exercised by the mind-brain of *Homo sapiens*, which is held constant. To an evolutionary bio-psychology, by contrast, genuine evolution of culture occurred only apace with the evolution of brain-mind from some level that characterizes infra-human primates, to whatever be the level achieved eventually by *Homo sapiens*. Holding constant the actual mechanism that evolves *ipso facto* precludes evolution of its product. The product can only elaborate.

1.2. I am compelled to ask your indulgence for a moment, since the previous statements appear to be critical before the fact. But the speaker has one foot each in the camps of bio- and cultural anthropology; thus criticism comes from inside. And this essay, too, has its biases. Its premises are 'organismal', since the thinking was shaped by that of Uexküll, of Weiss, of von Bertalanffy; therefore it is not atomistic or reductionistic, not neo-Darwinian, not 'behavioristic'. It will stress neuropsychology, a stress consistent with its systems-orientation. The papers that compose the substance of our Symposium, on the other hand, concern behavior. But I am convinced that sound behavioral analysis rests ultimately upon its neuropsychological fit; and I hope this sketchy attempt at a theoretic survey will convey to my colleagues my appreciation for what they are doing and will encourage them.

Some terms may now be defined:

Alloprimates. Any and all primates other than man. However, in this paper the implications shall be restricted to the monkeys and apes.

Phasia. The (human) speech function, a neuropsychological *ad hoc* programming.

Weiterbildung. There is no graceful equivalent in English. Succinctly, organic systems have evolved by generating further levels of order from antecedent, lower levels, a negative entropy, as developed by Schrödinger and others. In such a hierarchy, the next higher level is a new relational complex. This constitutes a *Weiterbildung*. The emergence of a brain with cerebral hemispheres from a prior level of a mere brain-stem kind of brain features a *Weiterbildung*. An evolution that simply differentiates without passing to a higher level is illustrated by the differentiation of two species where hitherto there was but one. *Weiterbildung* of neuropsychological system entails *Weiterbildung* of behavior.

System must be used in two senses: first, that meaning familiar from anatomy or physiology; and second, its meaning in current and general theories about the configurative relations of parts, one familiar from cybernetics, information theory, and so forth. In each case, the context should insure against ambiguity.

Mechanism has its meaning only within the second definition of 'system'.

1.3. *'Culture', sometimes so-called.* Let us dispose at the outset of what sometimes is called 'cultural', or at least 'intelligent', behavior among some invertebrates, particularly insects. Evolution of nervous control (as a lengthier discourse on the evolution of life-systems could show) should, on von Bertalanffy's principle of 'equifinality', eventuate repeatedly in processings of information we would term 'intelligent'. That lengthier discourse would also raise the puzzle: why did not at least several alloprimate lineages proceed to evolve culture, independently and in parallel? An idle question, of course; at all events, the australopithecoids became extinct, from quite unknown causes. The reason for mentioning the subject here is that we really are not justified in appraising the evolution of man's culturized life-mode as being an anomaly: it 'should' have happened repeatedly. 'Protoculture', I am ready to hazard, represents a normal and general level of life-mode for all higher primates, perhaps including even *Homo sapiens*. The presence of culture does not eliminate the possible enduring incidence of 'protoculture' as its matrix.

So we will assign to 'intelligent' behavior of insects an equifinality in the general evolutionary trend of nervous control, a systems matter. The evolution of behavior in the vertebrates raises deep problems of behavioral homology, problems we cannot solve as yet; for the evolution concerns neuropsychological processes, about which we indeed have information, but by far not enough. And we cannot transfer, *ohne Weiteres*, a definition of homology from comparative anatomy to comparative behavior. As we all know, behavior is a very fluid thing, amenable to permutations and to fresh combinings within different contexts. Yet we are justified in believing that behavioral homology exists.

1.4. The evolutional dimension of behavior confronts us with the truth that behavior does not really evolve. We may treat it, heuristically, *as if* it evolved, just as in evolutionary comparative anatomy we construct transformational trends from adult phenotypes. But 'behavior' has no independent ontology. It is what we see of the output from neuropsychological processings – their 'presenting symptoms', so to say. Moreover, in evolutional discourse, the minimum adequate universe is not the unit character but a *system*. This assertion is true simply because a whole is constituted of the information about the intrinsic properties of its parts plus the information about the relationships between parts; and this information is extrinsic to that of the parts. Organism-as-system evolves; parts adapt to the *organism;* it is the *organism* that adapts to the environment. This, of course, is no novel truth.

1.5. To the anthropologist in quest of evolutionary origins of culture, 'protoculture' translates as 'cultural anlagen' (disregarding here the possibility that something protocultural may have continued in man).

These identifications are intended to clarify and to orient what is to follow; yet, not unexpectedly, they uncover complications to any answering.

Consider a Japanese macaque washing a sweet potato and a Danubian washerwoman at work. The basic motor behaviors of both are on the same level of neuromuscular organization. And we cannot allow the environmentalist to tell us that the resemblances are purely 'learning'. All 'learning' *must* be channeled according to preexistent, 'predisposing' mechanisms capable of experiencing. However, this is not the main point here. Actually, macaque and human beings possess almost exactly the same motor repertoire, with but negligible discrepancies (the arm and hand of the macaque are, morphologically, not exact duplicates of the human organs). Similarly, the tyro pianist has the same fingerings as a Badura-Skoda. Both the Danubian washerwoman and Badura-Skoda, however, are vastly superior in their programmings of these limited motor resources. The cybernetist can explain the discrepancies in the same way as he would the expert bicyclist's ability to keep from falling off his machine. Experts develop a polarizing monitor which drastically narrows the ranges of a system of variables. Without this prerequisite, it is theoretically and practically impossible to program a series of motor patterns that eventuate in a complex series. But this concept introduces no new *levels of organization*. We shall explicate this point later on, following the late HENRY QUASTLER [1958]. Yet it is already obvious that when a mother mammal suckles her young, a feature totally absent in an alligator's or lizard's repertoire of maternalistic behavior has been introduced; and this ability represents a considerably higher level of organization. What, then, are the differences between the respective washings by Japanese macaque and Danubian woman? They are *extrinsic* factors; i.e., they are not built into the motor patterns themselves. It is in the extrinsic factors of the behaviors, numerous and not all of one class, that we must search for programmings that characterize, respectively, protoculture and culture.

This statement is almost a cliché. Yet it has its merits. Human beings at a zoo crowd for their most piquant entertainment before the monkey cages. What they are delighted to see is those creatures practicing the same basic motor behaviors they perform themselves; but the 'human' extrinsic factors are mostly nonexistent for the monkeys. Yet the situational contexts – a still more peripheral *Raumschale* or *emboîtement* – are there, roughly and transparently. Additionally, the performers themselves are grotesquely 'humanoid'.

The grand combination seems to reveal human nature bare-bottomed. To most, it is a laugh; but I have both lay and ethological friends who actually have a feeling of resentment toward these unabashed mockers of our own personalities. Well, at least we have, I daresay, the *hint* of a clue to the *Weiterbildung* of culture from protoculture.

Somehow, not only are the washerwoman and Badura-Skoda more 'skilled', as we have stated in cybernetic idiom; among the extrinsic factors are motivations of some higher order. Badura-Skoda *says* more than the tyro pianist does, or is even capable of conceiving. For the polarization of the variables, by narrowing their ranges, economizing energy and time, 'talks back' to Badura-Skoda. This is something much more than a feedback from a servomechanism. The pianist hears the playing as an experience; it acts as experiences do when they become parameters to a creativity that is autogenous within the organism. Now we are skirting the topic of 'sign' and of 'symbol', to which we shall return.

If we appear to be drifting away from a narrowly behavioristic definition of 'protoculture' toward, if we may use the word in some restricted sense, a 'mentalistic' one, this indeed is what we are doing. Certainly, we are not 'mechanistic'; definitely we are *mechanismal*, which is to be synonymized with organism-treated-as-system. For the *life-mode* of a human being, no matter what he or she may be doing, operates *always* on an identifiable cultural level, because the brain is what it is. The *life-mode* of an alloprimate, no matter what it may be doing, operates *always* on an identifiable protocultural level, because the brain is what it is.

1.6. Whatever be the case of culture, identification of protoculture begins with behavioral observation upon free-living alloprimates, no matter what has just been said. May I now be indulged the attempt to fit together a prospectus, beginning with the abstracts of my fellow-symposiants' papers before proceeding further? Soon we shall be hearing more about protocultural behavior in societies of Japanese macaques; group differentiae in the adaptive flexibility of sundry African monkey genera/species; particulars about the sex-bondings of hamadryas baboons viewed under a high-powered lens. Communication as messaging, furthermore, constitutes a particular topical dimension of social bonding; and isolated groups within the same species may develop their own viable dialects, as witness Japanese macaques. Particular notice is to be taken here of communicative changes with age. And along this particular dimension, we human beings are properly interested in anything alloprimate that suggests clues to the evolution of that special synthesis which is *uttered* thought: *phasia*. The synthesis is of course genuine enough, but thought and its utterance constitute no atomic phenomenon; and recently two small chimpanzees have been confounding those who have been stressing an ape's non-utterance to the neglect of its thinking. Art and play: whether and in what kind present in alloprimates, demand, of course, empirical ascertainment. Yet I daresay that in some future we shall come to understand that a thinking brain cannot but be an aesthetizing brain. Meanwhile how, under free-living conditions, do Washoe's and Sarah's conspecifics exercise in full chorus their protocultural powers? Finally, primate behavior is not to be understood until it has been viewed as a particular variety of mammalian behavior.

Every one of these far-ranging inductions tantalizes, for there seems to be some crypto-protoculture in each; what might it be? Let me attempt a suggestion. In respect to the social behaviors of beaver colonies, jackal packs, and monkey troops, it is the cognitive capacities of the latter plus the exercises of these capacities relative to what else the brain does that make their behaviors so much like our own, that lend their life-modes their 'humanoid style'. At present we must turn to the laboratory of primate psychology, rather than to the field, to find precisely the resemblances and differences between the cognitive processes of alloprimate and man. But it would be enjoyable to turn the Gardners and the van Lawicks loose together in a Tanzanian forest, or a Harlow and a Sade on Cayo Santiago, with the instruction to bring back to us a better idea of protoculture than we now have. And then, this bystander would beg a seat in a corner, while these investigators gathered around a low table with plenty of coffee.

Clearly, this argument is not to propose that cognition is all there is to protoculture. A mammal's behavior bespeaks the relatedness of its pre- and postcentral cerebral lobations: how both relate with the limbic system and how each and all of these structures relate with the brain-stem. So much for the individual. When two or more constitute a society, the latter has a 'style' gained from these individuals. It is for this reason that equating culture or protoculture with some transmissible learned behavior seems (to me) a facile procedure. Learning, after all, is merely something that happens to innately-determined neuropsychological constructions. You cannot characterize a plant by describing its flowers.

1.7. *Résumé;* a summary of our reasons for attempting a 'mentalistic' view of protoculture:

Behavior does not evolve; only (organic) systems do.

Behavior constitutes the 'presenting symptoms' of neuropsychological processes considered in their dynamic dimension. But we treat behavior, as a heuristic measure, 'as if' it evolved.

The distinctive structurings of these symptoms in the primates that eventuate in the 'primate personality' and, hence, in the 'style' of life-mode, we would specify as 'protocultural'. In this presentation, we shall allow 'personality' and 'style' of life-mode to be 'floating' terms; i.e., they will seek their own definition.

'Protoculture' is best treated *adjectivally*, before converting it to a nominal. This assumes that direct and extensive observations have already been made of alloprimates, so that their 'styles' begin to emerge from the episodal.

2. A 'Systems' Approach

2.0. But my résumé is more of a philosophical outline than of a scientific theory; we must try for the latter. Initially, one searches for a working definition of 'protoculture' by starting from culture and proceeding downward by a subtraction-and-contraction. The procedure soon reveals its limitations and the dangers of not observing them. We have noticed that there is no viable working definition of culture; those we have are logical considerations and do not identify organic processes. Logical categories of protoculture undergo no organic evolution into the logical categories of culture. This fundamental point could be made graphic by illustrating from the science of linguistics (in final analysis, a specialty-science within that of culture) by explicating that a system founded upon phoneme-morpheme-syntax does not lend itself to reconstructive speculation upon the organic emergence of phasia (some attempts to claim the contrary notwithstanding). But we will forgo the redundancy of such explication.

Furthermore, we will not tarry over the prescription that there must be transmittings of learning by way of interindividual communication if culture is to occur. This brevity is not intended to underestimate its importance. But to make of it a first-order requisite is to search for origins among *extrinsic factors*, not among *intrinsic features* and states nothing about intrinsic *neuro-psychological* properties that make transmission feasible, it states nothing about the kind of thing transmitted.

2.1. The evolution of nervous control, a very sound and substantial topic of neurophysiological systems and therewith of the 'as if' evolution of behavior, has followed a scheme of order-from-order hierarchy. Let us condense this subject into HENRY QUASTLER'S [1958] elementary statement from information theory that if we are dealing with a system of but one part, the problems are those of 'the efficient use of existing variations'; introduction of another part produces relations between parts; a third part introduces relations between relations; a fourth introduces relations between a part and a complex of relations. It is very noteworthy that each of these levels introduces a *quality* hitherto nonexistent.

Thus far QUASTLER. We should also note that complexity can be compounded further, if each of the parts can, on closer examination, also be resolved as a complex of relationships. Furthermore, parts may be indefinitely heterogeneous; relations between them are asymmetrical. In the brains of mammals, if a 'part' of a system is constituted of large populations of like cells (consider the layers of the cerebral neocortex), the brains may filter, transform, polarize, and relay the information coming to them from numerous sources. This processing is 'open-ended': theoretically, one can continue compounding *ad infinitum*.

Two very important properties are to be noticed. (1) In measure as we compound the relationships between constitutive variables (i. e., as we ascend the hierarchy), we rapidly gain further degrees of freedom of choices. (2) But the degrees are none the less channeled by the constraints that characterize the next lower level upon which the new synthesis has been built. Now let us move this abstraction toward the homely comparisons of alloprimates doing mammalian things in alloprimate ways, and human beings doing primate things in human ways. If we look down from man's height toward the primitive mammal, *plus ça change plus c'est la même chose;* the long view allows us to see *macrobehaviors* becoming regressively simpler, more restricted. Viewing the same thing from the bottom, primitive level to man, *plus c'est la même chose plus ça change;* the behaviors become progressively more complex.

A 'behavior' is always elected from a lattice of possibilities. Over an organism's lifetime, provided it is one of sufficient complexity, the organism never manages to exercise more than a fraction of these possibilities. And in the evolution of a system, it is not the behavior that is *exercised* that evolves, but the *possibilities*. Behavior is what is *observed to occur;* evolution is what happens to a *potential* for behavior. Once in a very improbable while, we may glimpse or sample the potential, as when Washoe or Sarah informs us, or when monkeys or apes seize immediately and avidly upon 'painting'. But Magdalenian man will never have the opportunity to demonstrate to us that he could have earned a Ph. D. A confirmed positivist probably would be less than happy with the notion that unrealized potentials must be taken seriously into account in a scientific treatise; nevertheless, they are being taken seriously in biological theory.

2.2. But these propositions about behavior must not degenerate into assertions of the 'who knows but that...?' variety. Our discourse must do better than this. It would be easy indeed to adopt KUMMER's beautiful little book, *Primate Societies* [1971] as a guidebook in the pursuit of protocultural behavior and to convert this paper into a marginal commentary out of systems theory. Surmisably, this type of commentary could also be done with the papers that follow this one; but it would not suit our objective. Let us begin differently.

No alloprimate ever communicates such precise and detailed information as bees do when reporting on the location of a pollen source. Are bees therefore more 'intelligent'? The bees possess a closed epistemological universe of discourse. They can communicate only about what 'everybody' is acquainted with. Their reportings merely place that knowledge in an 'interesting' context; the context itself is already a code. The 'surprise-value' resembles what we get from traffic-lights. We know that they are the only kind of messaging we are going to get at the next corner and that they will be red, yellow, or green. All we shall learn is the novelty of which it will be. But MENZEL's [1971] chimpanzees communicated to their companions that they should engage in an expedition of which nothing was common knowledge beyond the investigator's giving them the idea that something was worth an effort of goal-seeking. The chimpanzees' messaging has the far greater range of degrees of freedom. KUMMER [1971] states that 'primates have a potential for learning broad sets of tasks which neither they nor their ancestors encountered in this particular form. This flexibility, and not a specialized but genetically fixed skill, prepared the way for culture.' What KUMMER has seen in this descriptive way is being given more scientific precision by certain psychologists, notably by JEROME BRUNER [1969: 170] and his colleagues. From experimentation, he concludes that human infants mature their learning not so much as a specific response, but rather as a hierarchically organized, adaptive strategy of responses.

A more analytic discussion would show, I think, that an individual who cognizes thus strategically and within a greater range of degrees of freedom has a highly idiosyncratic 'personality'; and the *fabric* of its bondings with others supplies what we all term, vaguely, 'flexibility'. Does not all this suggest a canonical principle? Protoculture being a system of relationships, there must always be included in its considerations the highest level of order or synthesis obtaining in its hierarchy; and it must be assumed that this highest level is never abrogated. Conclusion: we must always seek the highest level.

2.3. In passing from a rather philosophical plane to one of scientific theorizing, we have come to see behavior as the manifest output of neuro-psychological mechanism. The 'as if' about behavioral evolution shows a successive buildup of orders of further processings applied to information and consistent programmings of performance. One may watch a mother lizard and a mother bird carrying through elaborate brooding programs whose features are, grossly, very much alike. Yet the bird's is much more embellished with detail; it sustains more *Umweg* procedures. Again, one may watch a rhesus macaque mother and a human mother carrying through elaborate infant-care programs whose features are, grossly, very much alike. Yet the human mother's is more embellished and sustains more *Umweg* procedures. Thus, our attention is drawn to macrobehavioral schemata, within which sub-systematic variations can occur. Let us consult neuropsychology a little more earnestly.

Between the superficial depth of the phenomena of observable behavior and the neurologic profundities out of which they must arise, there has to be logically an epigenetic zone of transition where the language of neurology is mapped by some topological transforms into the language of behavior; but it still is largely *terra incognita.* 'The language in which information is communicated [within the brain]... neither needs to be nor is apt to be built on the plan of those languages that men use towards one another' [PITTS and MCCULLOCH, quoted by PRIBRAM, 1971]. But these latter languages constitute behaviors, programmed *ad hoc;* they utilize exactly the same cerebral mechanisms as all other behaviors. In fact, the psychophysiologist HANS-LUKAS TEUBER [quoted by THORPE, 1969, p. 256] has said, 'Linguists are ethologists, working with man as their species for study, and ethologists are linguists, working with non-verbalizing species.'

It now becomes tempting to digress into the origins of phasia and to discuss concomitantly 'unbenanntes Denken' [KOEHLER, 1952], but time presses. Let us but continue from the observation that phasia in man cannot be a central topic here, because it concerns the *Weiterbildung* of culture from protoculture; and we are discussing protoculture. It is none the less of lively protocultural interest to determine what are the highest-level computings feasible in an ape's or other alloprimate's brain. Washoe and Sarah have demonstrated that they can be unexpectedly high. Of course, the chimpanzee's stubborn incapacity for using the instrument of vocalized code to express them is to us human beings a frustrating puzzle. But it would be salutary if for heuristic purposes we could dissociate verbalization from the gestalting of thought; for, after all, the two functions are not performed by the same parts of the brain.

Let us cut short this point of discussion. The capacity for thought is not dependent upon the capacity for putting it into vocalized code, even in man. That there should be some way and capability for programming an externalized behavior that is commensurate with the capacity for processing the inputs of information is hardly questionable. But something has gone awry when we hear (as we certainly do) the pronouncement that we start to speak as we think and end up by thinking as we speak. In view of everything that has been learned recently about the mentality of chimpanzees and their ways of manifesting it, we may well suppose that an evolved 'super-chimp' would possess the practical equivalent of our phasia, though put together not humanly, but 'chimply'.

2.4. *Cognitive processes in alloprimates*. We can hardly doubt that the mentations of mammals have a common, general neuropsychological base, which, if known, would make more intelligible the specifics of alloprimate and human thinking. We shall pay attention to the cognitive processes of alloprimates, those of macaques most particularly.

Here, as in all matters of alloprimate behavior, what the neuropsychological laboratory has to report should sharpen the observations of the primate ethologists. However, the job of the latter only begins *after* that; and what he finds is not of immediate practical application to the experimental neuropsychologist. Undoubtedly, the primate ethologist can appreciate that his monkeys resemble man in their processing of short- and long-term memory and that amnesias associated with hippocampal deficits travel in much the same way. However, it becomes a little more interesting to him, I imagine, to learn that man and alloprimates possess uniquely effective limbic systems in the amygdala and the hippocampus; that there is not much evolutionary difference in these systems between the human and the alloprimate; that they register and evaluate behavior [PRIBRAM, 1971], a process that requires a peculiarly effective short-term memory; that all these factors enter into the making of symbols, which is a capability of both alloprimate and man. Experiments indicate that man and monkey acquire habits and concepts in much the same ways. Concept learning is a matter of formation of learning sets. The learning of specific habits or the learning to discriminate novel qualities of objects, on the other hand, follows essentially other procedures. The experimental psychologist, DONALD MEYER [1971], is reasonably certain that in both human beings and alloprimates, prior habits interfere with the formation of new concepts. I, for one, doubt very much that the observed fact that female macaques learn from each other the trick of washing sweet potatoes, while the males ignore the procedure, indicates an intelligence differential; but, insofar as 'personality' may be considered an organization of habits, what may this behavioral difference suggest about the personality structures of males and females in particular societies?

Presumably, if some day analysis of the mentations of animals has become an integral part of understanding why they behave as they do, the primate-ethologist will also understand why the protoculture he observes cannot be otherwise than what it is, since it is performed by particular kinds of personalities. Like the anthropologist confronting human 'culturized' behavior, he will be interested in the *content* of the concepts, of *what* can be learned (and also of what can be forgotten), and the motivations involved. Non-verbalized concepting up to some level of complexity can be performed by birds, from the very intelligent raven to the stupid pigeon. Birds are capable of genuine counting and of making transfers of instruction from the somewhat arbitrary signs proffered them [KOEHLER, 1952]. But Sarah can manipulate a variety of discrete symbols to effect logical syntaxes.

Here it seems worth inserting that human nanocephals, with brains smaller than those of apes, learn to compose and autogenously to utter sentences from a verbal thesaurus, although, naturally, they fall far short of achieving the abstractive powers of the normal human being at early adolescence. Briefly, for the primate-ethologist, accounting for protoculture requires knowing not only *what* alloprimates can learn, but (very crucially) what is the *consistency* of their learning-to-learn, of their capacity for learning sets.

2.41. *'Body Schema'*. There are further dimensions to the problem. The human infant has the task of resolving his real world by polarizing it as an I and a non-I [KUBIE, 1953]. He tracks over his genetically predetermined schedule, and thus becomes progressively more capable of handling parameters (experiences). This formidable task eventuates, for one thing, in the assembling of what clinicians term 'body schema'; its existence became known to them from its bizarre distortions and disintegrations following lesions from rather well-defined cerebral locales. As the human infant becomes increasingly aware of the Out-There, his own toes (for example) first belong to it; gradually they are 'absorbed' and become attached to the Me. The Me goes on to acquire much more of a 'self', as we all know formally or otherwise.

How far down the vertebrate phylogenetic scale it might be feasible to seek the 'elements of ego', of course I do not know; how much it is the product of cerebral hemispheres is also an unknown. Yet we are certain, informally that monkeys and apes develop body schemata, as for instance, when they promptly put *our* spectacles over *their* eyes or identify *their* noses with *our* very different ones, to the point of using a handkerchief the way we do, Similarly, a chimpanzee knows what to do with your cigarette, and so on; but they do not behave thus toward other animals. There exists more precise evidence, however. Cortical ablations on macaques have been performed for years. DENNY-BROWN and his colleagues, for instance, ablated areas roughly homologous with those of human beings and produced behavioral disorientations strikingly reminiscent of human body-schema destructions.

We may be reasonably sure that protocultural behavior takes place between individuals having well-organized body schemata and 'selves'. Moreover, as we penetrate more deeply into the brains of monkeys and men, particularly to the limbic structures, the resemblances grow. This observation is not surprising, since there has been but little evolution between their respective limbic organizations. MACLEAN [1958] has characterized this portion of the brain as an 'illiterate' one, incapable of the fine-grained analyses of the neocortex; yet certainly it governs, shall we say, a 'contextuality' within which the neocortical executive must operate. In man, some limbic disturbances can be, symptomatically, very frightening. They have been known since at least the time of Hughlings Jackson and his case of 'Dr. Z'. MACLEAN [1958] has remarked that, but for the activities of these limbic mechanisms, 'we would seem much like disembodied spirits'.

2.42. The immediately foregoing remarks are a rather feeble attempt to underwrite protoculture as the result, not of what alloprimates learn, but of how they develop their strategies for learning to learn. Even a rat learns to learn. But the strategies and the content are different by as much as the architecture of the brain differs. For the moment, let us leave aside the neuropsychological mechanisms and confine ontogenesis of man and alloprimate strictly to behavioral considerations.

Recently, ALISON JOLLY [1972] has urged that 'we need real comparative developmental psychology, which will tell us not just what tests a cross-section of animals "succeed" or "fail" at but whether or in what circumstances they go through similar or different developmental stages as the human baby.' She notes further, that in certain tests, 'the strategies of the human baby, infant rhesus, and lemur are similar'. In fact, certain cognitive developments in rhesus macaque, chimpanzee, and man are sufficiently parallel as to suggest promise in stacking them against PIAGET's [1954] six levels of 'the ontogeny of logic'. She urges that (1) since his levels are 'natural', they are amenable to operations and (in the earlier of the human stages) are non-verbal, and (2) since his aims are not reductionistic, but seek to elucidate complex thought, his strategy could be used as a sort of manual for analogous studies upon alloprimates.

During the course of PIAGET's [1954] six 'stages', the child resolves and elaborates for himself a real universe and his own relation to it, along lines that proceed from sensorimotor accommodations to conceptions of objects, space, time, causality. Of course, it were impracticable, even if it were within your speaker's capacities, to review the body of researches into the ontogeneses of the logical processes in alloprimates and man; perhaps merely citing some names, such as LASHLEY, GOLDSTEIN, HARLOW, BRUNER, will be suggestive enough. It seems worth remarking, however, that although alloprimates parallel human beings in striking degrees over stretches of their respective developments, they drop out of the race far earlier than man and it remains uncertain just when and where they do so.

2.43. We have progressed to the point where 'sign' and 'symbol' demand consideration. And it is high time that a neuropsychology take over at the point where the foregoing discussions, which have uniformly tried to get along without it, have shown diminishing returns for their efforts. Thus, we should return to questions of functional localizations in the brain; but first a statement is in order about the paradoxes, better understood today than even a few years ago, of globality *versus* localization.

One of the fascinations of neuropsychology is its paradoxes, which, as the explorer meets their challenge by resolving them, produce further and quite as unexpected and even more intricate paradoxes. Computer models of the brain, Boolean algebra, and the various kinds of systems theory have proved powerful tools in coming to terms with the aspects of localization. And now comes holography. At present, it is not the physical structuring, but the functional implications that suggest models for the cerebral cortex and a resolution of the globality-localization paradox, although I am still incapable of explicating all this. Apparently however, even an EEG promises to tell more, and something different, than hitherto. At all events, we may be certain of one thing: the processes themselves must be the same in alloprimate and man.

The point to be emphasized now is that 'sign' and 'symbol' implicate respectively different hemispheral locations and that both are present in the alloprimates, as they are in man. Involved are what, in a now obsolescing terminology, have been known as the 'association' areas of the cerebral cortex and also some relations to the limbic structures. My remarks will consider 'sign' and 'symbol' in a neuropsychological framing only. The data do not support those philosophical-psychological treatments that would place symbol-making as some logical step that builds beyond sign-making.

Among mammalian cerebra, the primate is distinctive in the morphology and the degree of development of pre- and post-rolandic (pre- and post-central) 'association' cortices. Of their respective and distinctive functionings, it will be enough for our purposes to state very simplistically that the post-central, particularly the parietal, operates in analyses of spatial properties and of images and that the pre-central functions 'executively' in a kind of contextual, 'in-view-of-the-fact-that' way. What needs noticing is that monkeys possess, relative to total brain size, about the same amount of frontal lobation as man. In man, the relative amount of parietal cortex is a little, but not much, greater. The implication is rather obvious: monkeys must be capable of acting executively, or contextually, as effectively as man, within the limits of the informational analysis supplied from the rest of the brain involved. And we can expect from them a 'style' of life-mode that cannot obtain in carnivores, ungulates, Proboscidea, *et al*. This 'style' characterizes 'protoculture'. What that entity is remains to be found from inductions.

Now to hammer down the neurology of sign and of symbol to some extremely parsimonious statements. I am especially indebted to the neuro-psychologist, KARL PRIBRAM [1971], an imaginative, first-rank experimeter and a happy speculative genius, whose insights often prove to be so very right and are always seminal. His experimental subjects are mostly macaques, but his inferences to man are nevertheless professional. His experiments combine neurology and behavioral tests.

A 'sign' has a meaning free of context; a 'symbol' obtains its meaning from contexts that attach to it. This distinction holds, whether one is viewing the tokens of a 'chimpomat' or a swastika. Understandably, both sign and symbol handle 'imaging', so that a full argument would carry our steps yet farther back, to the problems of perception, which of course are multiple and multiplex. The elemental notion that a 'perception' is the working-over of sensory input is familiar enough; but a crucial extra factor must be added, one that is supported by good experimental evidence and theoretical rationale, that the *parietotemporal* 'association' cortex actively emits instructions to the perceptor (say, the retina) concerning *what to look for, on the premise of what the brain already knows*. This obviously promotes perception-cognition to a

servomechanism. To PRIBRAM [1971] a 'sign' results from the neuropsychological action upon imaging; clearly, 'contexts' do not participate in its
formation.

The mechanism that leads to symbol determination is distinctively different. The frontal cortex is continuous on the one hand with the (pre-rolandic)
motor cortex and on the other with the limbic formation; and the latter, as we
know, is the base of the temporal lobe. Experimental evidence indicates that
these structures are involved in converting presented situations into internal-
ized brain states. Sets of contexts are constructed from the deliveries of memory mechanisms that embody self-referrals [PRIBRAM, 1971]. To state the
difference between sign and symbol parsimoniously: the post-central association cortex operates on a continuum of perceptual images; the pre-central
association cortex imposes upon images a temporary organization of brain
events, i.e., context.

3. Toward a 'Mentalistic' Statement of Protoculture

3.0. Dr. MENZEL's earliest request was that this paper concern itself with
'protoculture' as 'anything bearing directly on group traditions, customs, and
inter-generational transfer of learned behavior regarding, for example, social
organization or communication; art, games, or play; predation and hunting;
travel paths and home-ranges; food habits; tool use; and so on'. Furthermore,
the Symposium, he wrote, would treat the more generalizing problem of man's
'uniqueness' among the primates and the distinctiveness of the primates among
the mammals in general. We must have continued our thinking along an
identical line; for shortly afterwards he wrote of the need for 'some very
critical thinking about whether, or in what sense, terms such as culture are
appropriate at all in animal research'. Well, I have been trying.

3.1. So now, within its limits, this sketch has led to the following deductions.

3.11. Under an evolutional premise, priority of definition belongs to
'protoculture'. Ways and means should be found for deriving 'culture' from
it, both theoretically and following operational inductions.

3.12. The *first* step in the strategy is, obviously, that of immediate observation of free-ranging primate behavior. But it should begin from *directed*

questions. It is understood that empirical data are not 'scientific fact' until their analysis has placed them in a context, which itself requires determining.

3.13. The laboratory of primate psychology furnishes a higher-powered lens and a narrowed field. Under the artificial conditions set up, the behavior of a monkey is about as 'protocultural' as that of a human being is 'cultural' under analogous circumstances. The ethologist's task begins where the experimentalist's leaves off. For the ethologist can know that the individuals he watches in the field are exercising mentations that perform as indicated by the experiments. This process parallels exactly the studies of man.

3.14. Behavior constitutes the *presenting symptoms* of neuropsychological processes. It is the *systems*, of which the processes are the *dynamics*, that actually evolve. The appearance of protoculture from antecedent absence of it presents an investigatory task; the appearance of culture from protoculture obeys the same principle. What a system *can* do is much more than what it ever does; it is the potential, rather than the realization, that evolves. The apparent dilemma, however, is not unresolvable.

3.15. A first-definition of 'protoculture' would state that it is the alloprimate 'style' of the general vertebrate-mammalian life-mode ('biogram'). At every turn, the behaviors of individuals are the outcome of neuropsychologies that have achieved a certain evolutional level. This sequence is the reason that our approach to protoculture is being termed 'mentalistic' and also that we are proposing that 'protoculture' should be treated adjectivally before being converted to a nominal.

3.2. Only two more tasks remain to this presentation. (1) It will be instructive to trace the evolutionary gathering of a complex that is both physical and behavioral from its lowliest anlagen, through its manifestation at an alloprimate level, to its eventual humanization (see 1a and 1b below). (2) Some specific sample suggestions from alloprimate behavior will serve to indicate where a mentalistic approach to protoculture may lead over to culture (see section 3.3).

(1a) This is a parsimonious sketch of the 'lactation complex'. There are 'unassembled' ingredients for it at the reptilian phylogenic level; an outline of its synthesis may be read (after a fashion) from proto- and metatherian organizations; it is complete in the eutherians; the mentalities of alloprimates have seized upon it to develop some very complicated mother-offspring relations; the mentality of man has carried the psychological *facies* even further, the physical *facies* hardly at all. The complex, of course, consists of a dyad, two heterogeneous organisms possessing a common genetic heritage but standing at two very different chronological states of development, plus the bondings between them.

At the reptilian level, maternalism includes both oviparity and viviparity and also, in oviparous reptiles, nest-building and tending of clutch; particular behaviors absent, of course, in viviparous reptiles. (But viviparous reptiles also furnish orientation to neonates.) The two kinds of partity occur in related genera; there is no reason to suppose that there is any corresponding discrepancy of intelligence. Some Lacertilia and Crocodilia are more comparable to each other in their maternalism than they are to other members of their respective orders. Viviparous Lacertilia manifest both yolk-sac and allantoic placentation.

Mammalia, but not reptilia, possess a peculiarly non-scaly, gland-ridden integument, underlaid by a muscular sheet that enshrouds the trunk from mouth to cloaca and innervated accordingly. The globality, i.e., the general homogeneity of the whole, is broken into by its locally specialized functions-and-structures: facial musculature that participates in a complicated oral mechanism; mammary glands, a further specialization from sebaceous glands; sphincters about the cloaca or the anus and about the region of the nipples.

Among reptiles, the snout is investigatory much more constantly than it is alimentary, as witness the nose and the tongue. Investigation is an organism's procedure for assembling the information for orienting the self. The mammalian *Weiterbildung* intercalates a suctorial period between birth and the supervention of self-dependence. The fetus pre-adapts by evolving lips and accessories and by complicating the tongue. Unlike the extra-embryonic membranes, the suctorial apparatus, although also an organ of immaturity, persists throughout adulthood; it develops secondary uses. But the self-orienting function of the snout persists, nevertheless, even in man, where a dramatic anatomical feature has been the loss of all but one of the mandibular bones and the structural and functional metamorphoses of the rest of these stem-ancestral, reptilian items into middle-ear ossicles.

With the evolution of suctorial apparatus goes an innate impulse to suck, a matter of brain-stem, under certain contextual conditions. Sucking is motorized from the fifth to twelfth cranial nerves. All of this elaboration is matched in the maternal system by mammary glands, with their autonomic innervation from the trunk and the impulse to suckle seated at least in some diencephalic-limbic composition.

The cardinal point is that all this development represents a logical assemblage of very disparate body parts and that nothing of it 'makes sense', except as referred to everything else, even to such an extent that the maternal evolution could not conceivably have occurred without the corresponding evolution in the offspring, and vice versa. This functional synchrony between asynchronic stages in the life-stages of two conspecific individuals is a quite arguable problem of holism, but is inarguable under reductionism.

I assume that my audience is well acquainted with the primate version of this mammalian basic scheme. Every order has its own diagnostic peculiarities; but set in the frame of QUASTLER's [1958] hierarchy of relational complexity, the variations constitute configurations of internal adaptations of systems all transpiring on the same level of complexity. Psychologically, the story is very different. In the mother-offspring bonding, the mentations of alloprimates exploit the physical resources they share with the rest of the mammals far beyond what other orders can do. As for man, the physical configuration is almost identical with those of the alloprimates; the diagnostic differences are hardly even secondary. For instance, the human and all alloprimate breasts are easily distinguishable by their respective shapes; yet I doubt that any one would be prepared to argue that their microstructures are at all dependent upon these shapes. But we must be very careful here, for the biology of form is a study as yet not beyond its infancy. If we consider the whole reproductive structure for the moment, perhaps the most striking difference between man and alloprimates is the presence in the former of the labia majora and their absence in the alloprimates. Their presence in man appears to be a feature of the distortion and squeezing of the perineum as part of the adaptation of this part of the body for orthograde stance and progression.

The severance of bondings between mother and offspring is a relatively very simple matter in reptiles. Of course, in any animal, the first step of severance, egg-laying or parturition, puts space between the two. In mammals, neonatally the physiological bonding is extended in lactation; but the psychological bonding is quite as real and is much more constant. Weaning severs the physiological bonding definitively, but the psychological one continues. In most mammalian orders, this severance is gradual but rapid; in the primates it lingers indefinitely. Dr. SADE has reported most interestingly on its reverberating among macaques even into adulthood; in man, it lingers perhaps throughout the lifetime. There may be a profound reason for this pattern. Recall the fact that at least as early as the reptiles, orality is investigative and hence self-orientative. *Nota bene:* no animal feeds, however hungry, unless it possesses first a degree of self-assurance from self-orientation. The human neonate cries, I believe, because it is lost. It will feed only after it has acquired a degree of environmental 'fitting'; I think every human mother knows this fact intuitively. Some years ago, HARLOW demonstrated it for macaque infants. The mouth-to-nipple 'syndrome' is a self-orienting focus even more forcefully than it is a feeding-device. But we must take note of the other member of the dyad. The mammary apparatus, as the psychoanalysts probably will tell us, embodies a focus of need in the human female. An adult female mammal *needs to be sucked.* In the human case, at all events, the 'need' perfuses the entire psyche. Its malfunctionings can be devastating, to her and also to her offspring. Is this phenomenon 'cultural'? No, obviously not. But in the human case, it *never* is expressed without cultural 'style'. Is it an agent that has participated in the evolutional generation of culture? My intuitive answer is yes. Is it at all implicated in 'protoculture'? Again, an intuitive yes. How? I would suggest: research the differences of mother-offspring relations in the higher alloprimates and the rest of the mammals; the field is virgin.

Culture is what human personalities do. This truth is being discovered by cultural anthropologists. It has given birth to an as yet uncertain domain, called 'personality in culture', that suffers, I believe, from its voluntary confinement to psychological foundations that remain biologically naive. Such a mistake could hardly be made in an inquiry into an 'alloprimate personality in protoculture'; someone should supply us with it.

(1b) Finally, in this matter, a theoretic note. In systems terms, the reciprocations between the heterogeneous moieties of a complex dyad are very involved and not yet completely grammarized. Still, if we take the cases of lactation or of copulation, either of the moieties involved may develop the psychophysiological 'need' to the level of motivational discharge ahead of the other. Reaching this level first may usually occur more frequently in the one moiety than in the other; e. g., we customarily think of a male initiative for the reproductive syndrome. We can begin to see this pattern as far down the vertebrate phylogenic scale as the fishes. At all events, whichever moiety arrives at motivation first messages a stimulus to the other.

3.3. In the search for a mentalistic definition of protoculture, one frequently encounters primate-ethologists' recordings of alloprimate behaviors that indeed remind one of human performances. They range from comparatively simple microbehavioral segments to rather lengthy stretches of macro-behavior. But the synthesis of highest possible order, which would relate even the latter into a configuration of life-mode, remains for the future, both for alloprimates and for man. Here are a few culled examples of alloprimate macrobehavior that make their bids to be considered protoculturally: baboon elders pausing during a trek to permit the youngsters to romp a while (WASH-BURN, DeVORE); a female baboon's tending the wounds of a male who has survived a serious fight (WASHBURN, DeVORE); a male baboon's 'supervising' a group of playing youngsters (DeVORE); greeting-embraces by adult female chimpanzees when they meet on a trail (the VAN LAWICKS); teamwork among adult male chimpanzees, as they try to capture a treed monkey (VAN LAWICK-GOODALL); an adult male chimpanzee's paternalistic interest in his mate's progeny [BINGHAM, 1927]; the 'festive' dance-playings by mixed chimpanzee groups (WOLFGANG KÖHLER, RENSCH); and chimpanzees' messaging to their group about *novel* environmental features [MENZEL, 1971].

These are high-order complexes indeed. Their significance lies in their open-endedness. On inspection, each could be shown, that is, to offer logical *Weiterbildung;* we will license ourselves to coin the metaphor of 'cultural promise'. The procedural hint comes from the kind of comparisons we en-countered when PIAGET [1954] was discussed earlier; we saw the human child's resolution of reality, which was both logical and biopsychological, and saw that alloprimates seemed to follow the same general logic, although not nearly as far. May I suggest that the key to understanding macrobehavior, of whatever level of order, lies in a specific and unambiguous conception of what is 'contextuality'. If this idea has ever been given serious theoretic attention, the literature is unknown to me.

3.4. *Homination.* Whether the specifically distinctive properties of 'cul-ture' belong in so elementary sketch of the idea of protoculture may be debatable. But if they serve to sharpen this idea, they become excusable. Our base is the open-endedness of the organic, evolving system that is the primate brain as a processer of information. The yet longer perspective is that of a vertebrate brain-stem into whose functional mechanisms have been inserted some highly abstractive ones termed cerebral hemispheres. These occur funda-mentally from monotreme to man, a by-no-means idle statement.

A dozen years ago, I drew attention to the human-culturization of an alloprimate sociality in these terms: males prolonged their associations with females to the point of developing an interest in the progeny of the latter and of mutual sharing of the products of both their skills; statuses became social generalizations; childness itself acquired such a social generalization. An integral consequence was the definition of elementary familialism ('nuclear' should be reserved for a later time, when extended relations developed). Human familialism receives sharper definition, as incest and homosexuality are recognized. Furthermore, it seems that among all vertebrates except man, inter-individual agonisms center more involved repertoires about the reproductive activities than in fields having no reproductive implications; in man, *per contra*, society becomes most complex outside the familial polarity. That these two 'societal' polarizations are not in reality unrelated, however, shows even in monkeys, where [as KUMMER, 1967, 1971, has noted] the male–female relationships vary ecologically.

The statements about sociality need psychological supplementing. The strategies of learning to learn and of what to learn (including what strategies to acquire) account, it seems to me, not only for man's particular recognitions of classes, but also for the way he classifies classes. Obviously, this procedure, too, has been an open-ended one, up to the limits of the mental capacity that exercises it. And this statement applies not only to objects but to the individuals around the classifiers; whence 'status' definition. A status represents a recognition by one's fellows. In human society, the classification of status receives further status: chiefs outrank commoners. The features of status are embodied as an 'image'; the image itself stands framed in, indeed is a part of, a social context. It seems to me very significant that although all alloprimate children play, only human ones play at being mothers, homemakers, hunters, soldiers, doctors, and nurses. They *evaluate adult statuses, and project their selves upon images of adult statuses. But they are similarly projecting adult statuses upon their selves.* And *nota bene:* a child can play at being a child, provided that, either, actually or imaginatively, another is playing at being an adult. *Homo ludens.*

These and other *excerpta* are interrelated in a multi-dimensional lattice. Notice, too, that they are all second, even third-order abstractions. Another line of inquiry would carry us from the goal seeking 'homeostats' and such, which are the serious playthings of cybernetists, to the summit-hierarchical levels of human values and the evaluation of values. And to reiterate: somewhere farther down on these ascents must dwell the various alloprimates. In principle, what goes for man-and-culture goes for alloprimate-and-protoculture.

Before leaving this highly rarefied atmosphere, please indulge me the surmisal that only an animal that can play at being an adult is also capable of asking itself throughout its lifetime the all-pervading question: *Who am I?*

The castings both of the question and of the infinitude of answers may range from a continuing, unsatisfied quest for an answer to a need for constant reassertion of whatever answer is found and a need to have others constantly redeclare the answer; also included may be a surrender of the quest. We call such surrender *schizophrenia*, although we usually think of it, I believe, as a flight from externality. Is it conceivable that, in some rudimentary degrees, the question begins at the protocultural level?

3.5. And a peroration. Protoculture is Janus-faced, in a way that culture is not. For behind protoculture lie its anlagen, represented in animals yet more primitive; while in front of it lie the *Weiterbildungen* that eventuated in culture. For culture, we are at liberty to seek the anlagen (in protoculture); but in front of it, thus far *ne plus ultra*.

That primatology, like anthropology, possesses a disciplinary frame of discourse, is evidenced by the entire program of this Congress. And *like anthropology*, primatology has not yet put together the image of its subject, to the point where it yields philosophical-scientific hypotheses from deductive principles. This presentation has tried to suggest the interdisciplinary schedule of resources indispensable to *both* our domains if we would achieve the further *étape*. Each of the disciplines mentioned already possesses bridgments to at least one other. We have a very long road ahead of us – but *what* a prospect!

References

BINGHAM, H.C.: Parental play of chimpanzees. J. Mammal. *8:* 77–89 (1927).
BRUNER, J.: On voluntary action and its hierarchical structures, see A. KOESTLER and J.R. SMYTHIES, eds. (Beyond Reductionism. Beacon, Boston 1969).
COUNT, E.W.: The lactation complex. Homo *18:* 38–54 (1967).
JOLLY, A.: The evolution of primate behavior (Macmillan, New York 1972).
KOEHLER, O.: Vom unbenannten Denken; in Verh. dtsch. zool. Ges., Freiburg, pp. 202–211 (Geest & Portig, Leipzig 1952).
KUBIE, L.S.: The distortion of the symbolic process in neurosis and psychosis. J. amer. psychoanal. Ass. *1:* 59–86 (1953).
KUMMER, H.: Tripartite relations in hamadryas baboons; in ALTMANN Social communication among primates, pp. 63–71 (University of Chicago Press, Chicago 1967).
KUMMER, H.: Primate societies (Aldine-Atherton, Chicago 1971).
MACLEAN, P.D.: Contrasting functions of limbic and neocortical systems and their relevance to psychophysiological aspects of medicine. Amer. J. Med. *25:* 611–625 (1958).
MENZEL, E.W.: Communication about the environment in a group of young chimpanzees. Folia primat. *15:* 220–232 (1971).

MEYER, D. H.: The habits and concepts of monkeys; in JARRARD Cognitive processes of non-human primates, pp. 83–102 (Academic Press, New York 1971).

PIAGET, J.: The construction of reality in the child (Basic Books, New York 1954).

PRIBRAM, K. H.: Languages of the brain (Prentice-Hall, Englewood Cliffs 1971).

QUASTLER, H.: A primer on information theory; in YOCKEY, PLATZMAN and QUASTLER Symposium on information theory in biology, 1956 (Pergamon Press, New York 1958).

THORPE, W. H.: in KOESTLER and SMYTHIES Beyond reductionism, pp. 428–434 (Hutchinson, New York; Beacon Press, Boston 1969).

Author's address: Dr. EARL W. COUNT, 2832 Webster Street, *Berkeley, CA 94705* (USA)

Symp. IVth Int. Congr. Primat., vol. 1: Precultural Primate Behavior,
pp. 26–50 (Karger, Basel 1973)

The Study of Infrahuman Culture in Japan

A Review

J. Itani and A. Nishimura

Laboratory of Physical Anthropology and Primate Research Institute, Kyoto University

Cultural Biology

Research on culture in wild Japanese monkeys had just begun in 1952, when Kinji Imanishi published a paper entitled 'Evolution of Humanity'. In August of that year we succeeded in provisionizing the monkeys in Koshima, Miyazaki Prefecture; and later, in November, a similar success was achieved with monkeys at Takasakiyama, Oita Prefecture [Itani and Tokuda, 1958; Itani, 1954]. Since research on Japanese monkeys was in the early stage of development at the time Imanishi published his paper, it is obvious that it was not based on detailed data about the ecology of Japanese monkeys.

'Evolution of Humanity' is written as a fantasy. That is, the paper takes the form of a panel discussion participated in by four characters – an evolutionist, a layman, a monkey, and a wasp – with the evolutionist as moderator. At the outset of the discussion, the evolutionist tells the other three panel members that an evolutionist's view of evolution is liable to be ignored, as is a priest's view of hell and paradise, and urges them to present their views so that, if any differences or similarities are noted, they might recognize them as such.

The first question taken up in this panel discussion is that of instinct and culture. Imanishi starts out the discussion with a comparison between instinct as the mode of living of animals and culture as the mode of life of man, saying that the former is genetic behavior while the latter is non-genetic or acquired behavior. Culture, he claims, must be based on group life, specifically a continued group life. Thus, a question arises regarding a group where behavior might be conveyed through direct or indirect contact between parents and infants, without such a mechanism as heredity in which germ cells are involved.

If there is a species that forms such a group, that group might be regarded as having its own culture, even if it were a group of wasps, according to IMANISHI.

In 1952, there were not much data available on this question, but IMANISHI made several specific forecasts. A monkey that had grown up in isolation from its mother since infancy might fail to show the normal behavior of monkeys raised in natural circumstances. Such an instance might reveal that what we have believed to be instinct is in fact culture. Some of the chimpanzees born and raised in the laboratory of R. M. YERKES [1943] failed to make nests for their beds; and some chimpanzees who gave birth to babies for the first time in the laboratory showed an awkward attitude toward their babies. Thus, IMANISHI maintains that such behavior might possibly be a form of culture, resulting from what the young animals learned from their group members, and further, that any difference in the preferred foods of groups of monkeys living in different environments might be considered as a difference in culture.

Before we use the word 'instinct' for a specific behavior, we must prove that it is not culture. Intelligence has been regarded as something opposite to instinct. There is no theoretical connection whatsoever between intelligence and instinct. It is impossible to define intelligence in terms of instinct or instinct in terms of intelligence. Instinct is synonymous with an inscrutable behavior. IMANISHI thinks of instinct from an historical point of view; he asserts that instinct is an inherited behavior and thus is something opposite to culture, which represents acquired behavior. If it is dogmatic to regard all animal behavior as instinctive, it is equally dogmatic to regard all human behavior as culture, says IMANISHI. Thus, IMANISHI restrictively uses the term 'adaptive behavior' for behavior that survived through natural selection and that is advantageous to the species. He uses the term 'adjustive behavior' for behavior that is acquired by learning. And he refers to culture as being adjustive behavior that exists under natural conditions.

Natural selection involves only elimination and selection. The prime force for evolution or creation is something different. If an animal lives only by instinct, we must consider an hypothesis for mutation. Animals which are capable of learning, on the other hand, will find a way of adjustment. Progress through adjustment of behavior transmitted generation after generation is not inconceivable.

IMANISHI uses the definition of personality as being an integrative system of behavior; he divides 'personality' into cultural personality and individual personality, and regards these as related to each other. He maintains that cultural personality serves as a basis of mutual communication to be shared by the group members and that individual personality is something that could bring forth a new cultural bud, depending on circumstances. Individual personality is similar to mutation as a hereditary phenomenon in that it is something creative; but if it were to become part of the culture of a group, various conditions must be met, as is also the case with mutation. IMANISHI assumes that individual personality is something acquired and at the same time is based on accidental, individualistic experience. The scope allowed for culture is the scope allowed for personality, according to IMANISHI.

From this standpoint, such expedient classifications as zoology and anthropology, based on discrimination between animals and man, become dubious. Either these two branches of science must come closer together, or something like cultural biology is necessary.

Comprehensive Study of Inter-Troop Behavioral Differences

IMANISHI's hypothesis was corroborated a few years after the publication of his 1952 paper by young students engaged in a field study of wild Japanese monkeys. The first report on that study was prepared by SHUNZO KAWAMURA: 'On a New Type of Feeding Habit which Developed in a Group of Wild Japanese Monkeys' [1954]. In it, KAWAMURA attempts to determine whether all the types of behavior he had observed in wild Japanese monkeys were what IMANISHI calls cultural behavior.

KAWAMURA first shows specific examples he obtained in the field regarding the repertoire of foods, nomadism of the troop, and feeding behaviors. In other words, KAWAMURA tries to confirm the presence of cultural behavior by checking on the variability of the behavior across troops. He gives some specific examples illustrating IMANISHI's assumption that there should be differences in food among different troops of monkeys and points out that young individuals seem to adjust to new foods much faster than older ones. This phenomenon will be discussed in detail later.

KAWAMURA's [1954] paper was the first report on the sweet potato washing behavior of the troop of monkeys on Koshima, which became well known later. This troop was provisionized in August 1952. In September 1953, a female monkey, F-111, about one year and three or four months old, was seen taking a sweet potato to a stream that runs through the island's sandy beach, washing the potato with water, and eating it. The monkey washed the sweet potato with both hands to remove sand from it so she could eat it more easily. In November of the same year, KAWAMURA conducted a study with MASAO KAWAI and found that a male monkey, M-10, who was one year older than

F-111, was washing sweet potatoes, too. Again in January 1954, KAWAMURA observed that another male, M-12, who was the same age as F-111, was also washing sweet potatoes[1]. F-105, mother of F-111, was observed to be very awkward at washing sweet potatoes at first but gradually became skillful at it, according to the report.

By the end of February 1954, there were only four individuals who had acquired the behavior of washing sweet potatoes. KAWAMURA assumes that the acquisition of this behavior by M-10 and M-12 can be attributed to opportunities for imitation provided through close social relationships with F-111 in the form of play. The fact that F-105 was the only adult monkey to have acquired this behavior is due to the close social relationship between mother and child. From these observations, KAWAMURA concludes that potato washing, which was initiated by a single monkey in the Koshima troop and imitated by others in the same troop, might be considered as a sort of culture that is gaining currency in the troop.

At the joint general meeting of the Anthropological Society of Nippon and the Japanese Society of Ethnology held in the autumn of 1955, IMANISHI and KAWAMURA spoke on the culture of wild Japanese monkeys[2]. Reactions to these speeches will be mentioned later. KAWAMURA made a summary report furnished with supplementary data in the next year's issue of *Shizen* magazine [KAWAMURA, 1956]. At that time, field research on Japanese monkeys was well under way; and, in addition to the data obtained from continuous observation of the Koshima and Takasakiyama troops, a fair collection of comparative data on more than ten troops, including Minoo, Toimisaki, Arashiyama, Shodoshima, and Taishakukyo, was made available, so that KAWAMURA was confident of the presence of culture.

For example, one troop may ravage a certain farm crop in the area adjacent to its nomadic range; but another troop does not. KAWAMURA also observed that some troops liked eggs, but some did not. The paternal care observed in the Takasakiyama troop [ITANI, 1954, 1959] was seen in only a few other troops. The male-male mounting behavior that was frequently observed in the Koshima and Takasakiyama troops was not seen in either Troop-S or Troop-K on Shodoshima. Instead, the most dominant male was reported to have the behavior of ritualistic mounting of females. HIROKI MIZUHARA [1957] reported several characteristic sexual behaviors of the

1   This individual was later removed from KAWAMURA's list [1956]. KAWAI [1965] maintained that M-12 did not wash sweet potatoes in 1962.
2   KAWAMURA's speech was titled 'On the Cultural Phenomena in the Society of Wild Japanese Monkeys'.

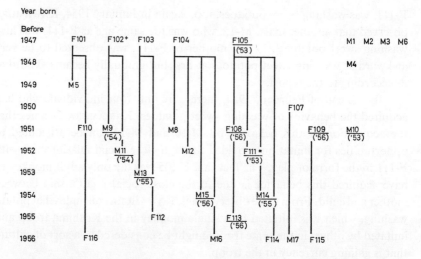

Fig. 1. The lineage of Koshima troop and the diffusion of sweet potato washing behavior [after KAWAMURA, 1956]. + = Died; * = originator; – = sweet potato washing monkey; ('53) = the monkey began sweet potato washing behavior in 1953.

Taishakukyo troop. This troop showed no signs of masturbation, which is commonly seen among male monkeys in other troops. In the context of the male monkeys' display toward females, the peculiar behavior of a male pulling the chin of a female was observed. And another behavior not seen in any other troops, that of a male pulling up the tail of a female and peeping at her genitals, was observed.

By this time, the behavior of washing sweet potatoes had spread to 11 of the 30 monkeys that constituted the Koshima troop. Figure 1 shows the genealogical relationship of this troop, with the trends of diffusion of this behavior. Although it is difficult to read definite trends from this figure, it can be assumed that a new pattern of culture is spreading to the younger generation and to kinship groups through maternal lines. The originator of this behavior, F-111, invented another new behavior. She gathered wheat that had been scattered on the sand. Scooping up the wheat in her hands, she took it to the water and threw it into the water to separate the wheat from the sand. The spread of wheat-washing behavior was followed by KENJI YOSHIBA, SHIGERU AZUMA, and KAWAI in subsequent years [KAWAMURA, 1956; KAWAI, 1965].

These are the kinds of culture that can be confirmed visually by observers. KAWAMURA calls them 'overt cultures' and tries to find in them evidence of 'implicit culture'. From his experience in the provisionization of many troops of monkeys, he points out that there is a considerable difference in the ease or difficulty of provisionization among different troops, even when the same methods are used. Normally, monkeys are provisionized after a few months. But Troop-A at Minoo was provisionized in just a week; whereas the Takagoyama troop has not yet been provisionized after a year and a half. There is also a considerable difference among various troops in the speed of acquiring new food habits. KAWAMURA considered that the basic attitude of each troop of monkeys was supported by the social organization of that troop and that this basic attitude was the most important thing that individual monkeys born to that troop must learn and acquire. Thus, KAWAMURA tries to find in a troop the presence of that integrative culture or 'personality' that IMANISHI assumed to be the integrative system of behavior.

These early field surveys and reports revealed the differences that existed between troops in their behavior and attitude and thus challenged the instinct theory. Because his report was based on field observations, KAWAMURA found it difficult to devise an entirely satisfactory scale of comparison. Although some of his remarks had to be corrected later on, his statement was enough to corroborate IMANISHI's hypothesis as to the presence of cultures.

KAWAMURA's speech aroused a heated discussion and even derision over the question of whether or not monkeys have culture. A few students, however, decided that they would carefully see to the future development of his study. Following the publication of a report by KAWAMURA [1956] entitled 'The Prehuman Culture' and another by ITANI [1956a] entitled 'The Prehuman Language' in one issue of *Shizen*, the same magazine carried an article in the next issue which gave the opinions of seven scholars on the two reports[3]. These seven scholars were well-known zoologists, psychologists, philosophers, sociologists, and cultural anthropologists. Three of them raised questions regarding methodology, the results, and the way of thinking; another three reserved judgment and stated that they would expect further studies in the future; and the last criticized the work by saying that it would be like putting the cart before the ox to consider that the day would come when studies of Japanese monkeys would help define the culture peculiar to man. By 1959, KAWAMURA came to use the word 'subculture' because he thought it would be wise to avoid unnecessary argument. The subject of whether or not monkeys have culture became less and less discussed, and no definite conclusion was reached. A few years after the first big debates, an outline of Japanese research on the culture of Japanese monkeys was produced in English in papers by

3   Seven views on the life of monkeys, *Shizen 11 (12)*: 2–10 (1956).

IMANISHI [1957b, 1960a], DENZABURO MIYADI [1959], JEAN FRISCH [1959], and
KAWAMURA [1959, published in 1962]; and thus it came to be known at long
last among foreign researchers. As long ago as 1954, however, a 16 mm film
entitled 'Natural Society of Japanese Monkeys' was produced; it included a
scene showing the early potato-washing behavior of monkeys in the Koshima
troop.

A comprehensive study on subculture, based on a comparative study of
various troops and large amounts of data, was compiled into a single report by
KAWAMURA [1965]. In defining culture, KAWAMURA uses the definitions of
C. KLUCKHOHN and W. H. KELLY [1945] and R. LINTON [1945], which exclude
such factors as symbolic function, tool using ability, and productive activity.
However, whereas KLUCKHOHN limits the term culture to human activities,
KAWAMURA maintains that these definitions could be applied to the cultural
phenomena of higher animals without contradictions.

Taking note of the concept of 'canalization' proposed by KLUCKHOHN,
KAWAMURA states that an instinctive act requires no canalization because it
would naturally progress along a specific course and that canalization would
be required only in cases where many alternative courses are possible. That is,
KAWAMURA deals with culture as social canalization from the standpoint of
function and maintains that the essential quality of subculture lies in the
averaging of many possibilities into a uniform behavior through canalizing
them into a basic canal.

At the time when KAWAMURA's [1965] paper was published, about 20
troops of Japanese monkeys were under study; that KAWAMURA was himself
studying 11 troops tells how hard he worked in comparing different troops of
monkeys. His report covered so comprehensive a range in the comparative
study of behavior in Japanese monkeys that one might suspect that he was
talking about pan-cultural theory. To avoid repetition of statements made in
the foregoing pages, no details of KAWAMURA's 1965 report will be given here;
but it might be noted that it is more complete than KAWAMURA's preceding
papers. He points out, in the final analysis, the importance of understanding
the presence of a more wholistic and integrated troop culture and the mechanism of its basic canalization.

As an effective means of analysis, he planned to trace the process of
development of a number of infants who had been isolated from their mothers
soon after birth, to gather them into an artificial group after a few years, and
to observe the culture developed by them. SUMIKO KAWABE [1964, 1965] took
care of the observation of the isolated monkeys; and in 1965, 11 monkeys
whose life histories had been recorded in detail were released on Tsurushima,

an island in the southeastern corner of the Straits of Kitan. The strange food habits of this artificial group were reported by SHINGO MAEKAWA [1967]. However, it is greatly regretted that the unexpected death of SUMIKO KAWABE in the spring of 1972 has prevented further reports on this work.

Propagation and Tradition in the Takasakiyama Troop

Along with the descriptive studies mentioned above, ITANI [1956a, 1958] proceeded with an analytical study of the culture of the Takasakiyama troop in 1954. The purpose of his study was to analyze quickly the origin of sweet-potato-washing behavior, which was in a very slow process of diffusion among the monkeys of the Koshima troop.

ITANI [1956b] notes that the monkeys at Takasakiyama consume about 120 kinds of wild plants during the year, and he has a strong impression that there is a definite distinction between those potential foods that are recognized as such by the monkeys and those that are not. For example, the monkeys readily eat the leaves and fleshy buds of *Dioscorea japonica*, but show no interest at all in its root stocks after they are dug out of the ground. The Takasakiyama troop knew sweet potatoes and wheat before they were provisionized because they ravaged nearby farms for these foods; but in the early days after provisionization, they showed no interest in the apples, tangerines, and soybeans that were thrown before them. Although they had come to eat these foods by 1954, they would not touch artificial foods offered by visitors.

Candy was selected as a new test food. It was expected that the monkeys would take a long time to accept the candy as a new food. However, it was considered likely that once it had been tasted, candy would become the monkeys' favorite food. The test was designed to determine age and sex differences in the ability to acquire a new food habit; to trace the route of propagation (if the habit of eating candy spread from one individual to another); to study the relationship between habit propagation and social organization; and to consider the communication systems used in habit propagation.

In May 1953, only one adult male, Achilles, was observed eating a candy; and no other individuals were seen eating any candy. The test was conducted a total of six times during 14 months beginning in July 1954. By offering candies to each of the monkeys who were individually known to him, the investigator traced the process of acquisition in these selected individuals. The

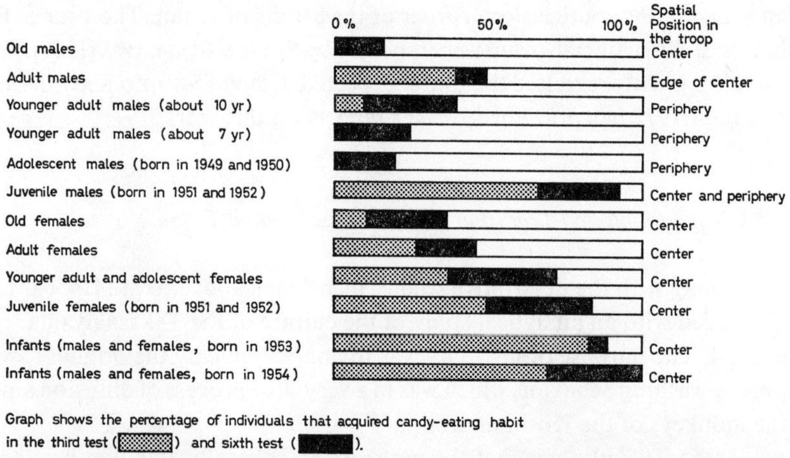

Graph shows the percentage of individuals that acquired candy-eating habit
in the third test (▦▦▦▦) and sixth test (███).

Fig. 2. Acquisition process of candy eating habit in Takasakiyama troop [after ITANI, 1958].

troop size was about 320 at the end of the breeding season (May–August) of 1954 and had increased to about 370 when the sixth test was conducted in September 1955.

Figure 2 compares the rates of acquisition in different age groups, using the data obtained from the third and sixth tests. The column at the right shows the spatial positions held by the respective age groups in the troop. As is clear from the figure, the young monkeys showed a more positive attitude toward candy than their elders and a higher rate of acquisition. The elder monkeys were conservative toward things new. The females generally showed a higher rate of acquisition than the males. This pattern indicates that the propagation of this behavior resulted not only from 'invention' based on the trial and error of individual monkeys but also from the 'imitation' of those who had already acquired the behavior by those who had not. In other words, female monkeys with infants maintained close social relationships and were influenced by those infants who showed high acquiring ability. The same can be said of the males. Although the leader males showed a low rate of acquisition due to old age, the sub-leader males showed a relatively high rate of acquisition, conceivably due to the paternal care [ITANI, 1956a, 1959] shown by the males of the sub-leader class and above. On the other hand, the younger males who had no habit of paternal care were left out of the propagation of this culture, as they lived in the peripheral part of the troop.

It is worth noting that individuals 1–3 years of age showed the highest rate of acquisition and almost without exception acquired the behavior during the test period. ITANI considered that this showed their strong interest in things new and in the behaviors of other members of the troop and, therefore, turned his attention to the play group which served as a pool for propagation. As compared with infant and juvenile monkeys, adults undoubtedly acquire more by imitation than by individual discovery; and the rates of acquisition of various age groups of adult males (4 years or older) vary depending on their position in their troop – that is, by the degree of contact with infants.

As the candy tests indicate, a new behavior is acquired by the younger generation of a troop; and, if it is useful for them, it will be propagated to the older members of the troop through close social relationships and become a new habit, or a new culture, of the troop. Propagation among playmates, from infants to mothers and among brothers and sisters, and from infants to their male protectors, must be the most important channels of introducing new forms of culture.

Individuals born in 1954 were born about the time that the first and second tests were conducted. When the third test was made about a half year later, 70% of the monkeys had already acquired the habit of eating candies. But it must be noted that before their acquisition rate had reached 100%, by the sixth test, propagation in the reverse direction (that is, propagation from mothers who had acquired the habit of eating candies, to their infants) helped the infants acquire the habit. In this case, however, the propagation speed must have been great because it was the infants that were learning the habit. These findings have a close relationship to the results of a survey by ITANI [1956b] of the natural foods consumed by monkeys at Takasakiyama. His results indicate that infants seemed to acquire practically all foods they were ever to be allowed to take during their first year of life, when they were most heavily dependent on their mothers. The old culture of a troop must of course be transmitted from older ones to younger ones in the reverse direction of the channel mentioned in connection with the acquisition of new habits. ITANI called this 'reverse' process 'tradition', as opposed to 'propagation'. He further states that it is one-sided communication, dependent upon the ability of communicatees, and that he has found no vocal sounds or gestures among communicaters and communicatees which would accelerate diffusion.

From the fact that solitary males join other troops, as revealed by MUNEMI (KAWABE) YAMADA [1966], ITANI also reasoned that it was possible for solitary males to propagate their own culture to other troops. But in view of the fact that there were many adult male monkeys who still would not eat

candies 26 months after the start of the candy test and that their culture would
be propagated only through close social relationships, ITANI maintains that it
must take a long time for the propagation of culture from one troop of
Japanese monkeys to another.

Propagation and Tradition in the Minoo Troop-B

The year after ITANI finished his test, M. (KAWABE) YAMADA [1957] traced
the course of diffusion of wheat eating in the Minoo Troop-B. His results
showed a striking contrast to those of ITANI. An outline of YAMADA's results
will be given briefly in this section, and the difference between these two tests
will also be discussed.

At the time YAMADA's test was conducted (March 28–29, 1956), the troop
was a small one comprising 17 individuals: two adult males, five adult females,
three young females, four juveniles and three infants. Among them, only
Nasio, the second dominant male, had once left the troop in the past and
acquired the habit of eating wheat. YAMADA took this instance as a case in
which, contrary to the candy test at Takasakiyama, a principal member of
the troop rather than a juvenile might act as the instigator of habit diffusion.
He commenced the test by arranging two shallow wooden boxes containing
wheat at intervals of 10 m on the feeding ground.

The results for 14 individuals, except three yearlings, are shown in figure 3.
Only Nasio was eating at the beginning, but several minutes later Ata, the
most dominant male, became interested in what Nasio was doing and started
to eat also. Similarly, an hour after the wheat was provided, Zuku (a dominant
female) and her infant Tanu took part. These four individuals were then
followed by Yami, a female in the central part, and by Anzu, a 4.5-year-old
daughter of Zuku, making a total of six individuals. In three hours, Nobara, a
female in the central part, Buna, a peripheral female, and her 3.5-year-old
juvenile, Itigo, had also participated, followed by a 2.5-year-old female, Nemu,
and a 1.5-year-old female, Lulu (the juvenile of Yami), a total of 11 of 14
individuals thus far coming to eat wheat. Another hour later, the remaining
peripheral females (Kaede and her two daughters, 4.5 and 1.5-years-old)
started eating; thus, the diffusion of wheat eating across the entire troop was
completed in the short time of only four hours.

The characteristic features of this diffusion are: (1) it was much faster
than the diffusion of potato washing and candy; (2) in each lineage, the habit
was diffused starting with principal individuals in the central part of the

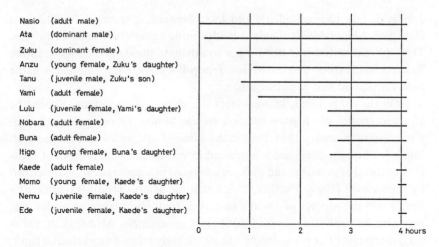

Nasio	(adult male)
Ata	(dominant male)
Zuku	(dominant female)
Anzu	(young female, Zuku's daughter)
Tanu	(juvenile male, Zuku's son)
Yami	(adult female)
Lulu	(juvenile female, Yami's daughter)
Nobara	(adult female)
Buna	(adult female)
Itigo	(young female, Buna's daughter)
Kaede	(adult female)
Momo	(young female, Kaede's daughter)
Nemu	(juvenile female, Kaede's daughter)
Ede	(juvenile female, Kaede's daughter)

Fig. 3. Process of the acquisition of wheat-eating behavior in Minoo Troop-B [after YAMADA (KAWABE), 1957].

troop; and (3) as a whole, the older individuals were earlier and the younger ones later in acquisition. In the case of Zuku's lineage, first the mother Zuku acquired the habit, and the diffusion took place from her to her juveniles, first to Tanu and then to Anzu, within only 30 minutes. And the time between Itigo's acquisition and her mother Buna's was 15 minutes at the most. Three individuals in the Kaede lineage, exclusive of Nemu, were the last; but they acquired the habit almost simultaneously a little before the end of the fourth hour. Examining this information more closely, we see that rather than propagation passing from elder to younger monkeys as described by YAMADA, propagation took place first among five individuals, from Nasio to Yami, who formed the central part of the group. After that propagation within lineage played a major role, the details of which include not only propagation from elder to younger monkeys, but simultaneous propagation in the opposite direction.

Examining the cause of the difference from the diffusion of candy eating at Takasakiyama, YAMADA indicated the following points. First of all, propagation from elder to younger monkeys was undoubtedly an important cause of an accelerated propagation. He contends that a difference in social organization between the Takasakiyama troop (advanced in status differentiation) and the Minoo Troop-B (undifferentiated in that respect) should be taken into account. He cites an instance where candy diffusion, similar to that in the Takasakiyama troop, could be seen in the Shodoshima Troop-K, comprising approximately 130 individuals and less advanced in status differentiation; almost all the individuals acquired candy during a period of about six months. Thus, he states that it is not proper to attribute the difference between the Takasakiyama troop and the Minoo Troop-B merely to the differences in the food introduced and population size.

On the other hand, ITANI adopts the view that importance should be attached to the life history of each troop. Minoo Troop-B, though provisionized in February 1954, lives in the Minoo Park, which is thronged with tourists. These monkeys are accustomed to various artificial things not existing in natural mountains; and they are advanced in 'canalization', as described by KAWAMURA [1965]. Further, a view that the Minoo Troop-B represents a case where the process of ITANI's so-called 'tradition' played an active role may also be applicable. ITANI [1958] gives an example. Monkeys in Takasakiyama, soon after provisionization, obstinately refused any artificial food; but an infant who was lost and was then taken under human protection started to take any artificial foods offered in only an hour. If it is assumed that group life has some power to inhibit individual monkeys living in the troop from acting at their pleasure and that such an inhibitive force is their culture itself, isolation from the group life will mean a release from such inhibition and, accordingly, from a certain traditional culture. If one takes such a viewpoint, the relation between cultural pressure exerted upon individuals and population size may not be disregarded; and further it is necessary to reconsider the behavior of solitary males leaving the troop.

Identification Theory

IMANISHI had continued extending his thoughts on 'the behavior brought to a complete adjustment in natural conditions', partly through reviewing the data collected by young researchers in the field and partly through other materials. In 1957, he introduced in both Japanese and English the outline of our studies up to that time on the Japanese monkeys [IMANISHI, 1957a, 1957b]; but not much space is spared therein for culture. However, in the same year he wrote two articles [IMANISHI, 1957c, 1957d] that one may say are groping for basic theories to support culture.

If many of the behaviors exhibited by wild monkeys are not instinctive, but cultural, in what manner are these behaviors learned? Speaking of a primitive culture released from the ties of instinct, one is apt to have an image of something weak in vitality. However, the monkey troops one observes in the field are full of vitality, having a splendidly complete behavior system. IMANISHI classifies their behaviors into 'individual-oriented behavior' and 'troop-oriented behavior'. He considers that as long as even the individual-oriented behavior of eating one thing and not eating another is endorsed by a culture originating in each troop, it is not acquired merely by individual trial and error. Beyond these behaviors, on the other hand, in seeking a principle to account for both types, IMANISHI brought such behaviors as dominance rank order and leadership face to face with his own theory of culture.

Among the troop-oriented behaviors, there are some in which individuals are not rewarded for imitating others. First, following the learning theory of O. H. MOWRER, IMANISHI [1957c] tried to examine the learning process of monkeys from two aspects of conditioning and problem solving. He used the 'warning sound'[4] [ITANI and HAZAMA, 1953; ITANI, 1954, 1956a, 1963] as the basis for his discussion. A tense and keen voice is raised by only one individual (a male in most cases) if an invader approaches the troop, and the individual giving the sound stands against the enemy alone and repeats the warning sound. Meanwhile other individuals hearing it escape in a group from the site. Though such a sound is usually given by the most dominant male among the monkeys perceiving the invader, if a juvenile should be first to give the alarm, the warning may be relayed by the stronger voice of a more dominant individual in the troop.

Though the details of analysis tried by IMANISHI [1957c] in accordance with the learning theory of MOWRER will be omitted here, in conclusion he states:

'The experiments of animals in the past have all been individual experiments, eluci-dating accordingly individual-oriented behaviors. Therefore, reward was required at any time. However, the troop-oriented behavior sometimes conflicts with the individual-oriented behavior. If it should be done without a reward before one or if any, by disregarding it and is so trained as to be done, its learning may be said to be learning to introduce superego and learning to acquire culture.'

In that paper, such a social role as the utterance of warning sounds and also the recognition of ranking and social position, etc., are all treated as learning; the same subject is further discussed in his next paper [IMANISHI, 1957d].

IMANISHI believes that a mother monkey must play an important role in her infant's learning various aspects of culture in the troop to which it belongs. The infant obeys the mother, imitates what she does, and learns to eat what she eats; this process can be regarded as a matched-dependent behavior, as

4   D-group sounds and in particular, D-1, according to the classification of ITANI. [1963]

described by MILLER and DOLLARD [1941]. Similarly, from behaviors and
attitudes taken by the mother in various group situations, the infant deepens
its recognition of individual members in the troop and gradually acquires
behaviors and attitudes in the manner of the mother. Social rankings, too, are
learned as one of the attributes inherent in the troop without the process of
trial and error. At the same time, if there are leader males present to control the
troop and if the infant's mother always behaves as a follower with respect to
the leader males, how would leadership of these leaders and the troop-
oriented behavior be conveyed to the infant? The matched-dependent behavior theory cannot answer this question. If a young male behaves in the
fashion of a nearby leader, it is imitation. If, however, there is no leader to
follow or imitate and the young male should do something that would
ordinarily have been done by a leader, it cannot be said to be imitation. Here
IMANISHI [1957d] applies as the explanatory principle the identification theory
in psychoanalysis. 'Identification' as referred to here is nothing less than
introducing one personality into another; the personality thus introduced is
superego. The realization of identification depends on the success of MOWRER'S
so-called 'conditioning', but at the same time, 'problem solving' should accompany it.

Though an infant first takes the mother as the object of identification, in
the case of a male monkey (as in human males) the time must come when the
infant will transfer the object of identification to dominant males in the troop.
Thus, IMANISHI tries to re-explain, in terms of identification, the behavior of a
young male giving forth a warning sound for the first time to warn against the
approach of an invader. However, unlike the case of learning from the mother,
the object of identification cannot be just any male. In accordance with the
theory of MOWRER [1950], IMANISHI contends that the second object of
identification for males should be a leader whose status and role correspond
to those of a father in the human family. However, there arises here the problem that identification by infant males will not always go right. At this point,
IMANISHI takes up the question of 'breeding'.

Though not made fully clear at that time, it was predicted that the central
part of a troop comprised a number of matrilineal groups[5]. Further, it was
known that ranking existed among the females, too, and that dominant
females in the central part of a troop maintained a closer contact with male
leaders. Although it was known that in the Takasakiyama troop during the

5   The structure of the matrilineal genealogical group was later made clear in detail by
NAOKI KOYAMA. [1967]

breeding season, leader and sub-leader males took care of the infants born in the preceding year [ITANI, 1959], it was assumed that since the infants of more dominant females were always near male leaders, there might be some infants that were protected by the higher ranked leaders and that, therefore, acquired a more complete identification with them. Though the period in which leaders take care of infants is two to three months at the most, the attitude implanted in infants during that period probably remains with them into the future.

ITANI [1957] reported that despite being 5 years old, Toku, the son of the dominant female, Til, in the Takasakiyama troop, did not move to a peripheral part of the troop and continued staying in the central part, where he was also tolerated by females. IMANISHI [1957d] suggested that this might be a case in which identification worked well. From the outline given of IMANISHI's identification theory, one can understand how difficult it is to prove objectively, on the basis of observation, the process by which one personality is introduced into another personality. IMANISHI comments on the process as follows: 'We cannot know in what manner the present leader males have become the leaders. However, at present where the troop has been provisionized, by tracing the life history of infants born in the troop it must be possible to verify the hypothesis set up in this paper.' That is, he set up one of the guiding principles in a long-term diachronic study.

In 1960, IMANISHI tried to generalize his identification theory, and discussed the relationship between identification and imprinting in birds [LORENZ, 1952]. He made clear the difference between his viewpoint and that of FREUD [1933], but the particulars of this discussion will not be covered here [IMANISHI, 1960b].

Genealogy and Culture in the Koshima Troop

The behavior of sweet potato washing in the Koshima troop has slowly but steadily been diffused since 1953. The number of individuals washing sweet potatoes increased from 11 in 1956 to 15 in 1957 and to 17 in 1958. By 1962, the size of the troop reached 59 individuals, among which 36, or 73.4% of monkeys older than one year, had acquired this behavior. Similarly, the wheat-washing behavior commenced by F-111, and first observed by KAWAMURA [1956], was performed by eight individuals in 1959; and by summer 1962, an additional 11 individuals (or a total of 19) had taken up this behavior. [KAWAI, 1964, 1965]

In the period from 1960 to 1962, KAWAI [1964, 1965] made an intensive survey of cultural behavior, concentrating especially on the Koshima troop. He attempted an analysis of the process of diffusion of the two behaviors mentioned above and two other new behaviors, such as 'bathing behavior', in which monkeys bathe in sea water, and 'give-me-some behavior', in which monkeys hold out their hands toward persons visiting the island to ask for food, etc.; these latter two behaviors were then spreading through the island. He used the term 'pre-culture' for the behaviors called 'culture' by IMANISHI [1952] and 'sub-culture' by KAWAMURA [1959]. It is deemed to be a designation with an evolutional intent.

KAWAI divided the diffusion of the sweet potato washing behavior into two periods, designating the period from 1953 to 1957 as the first or 'the period of individual propagation'. The general characteristics observed in that period were: (1) that young individuals (1–3 years old) showed a high rate of acquisition; (2) that propagation from younger to elder individuals could be observed; and (3) that a sex difference in acquisition was noticed, that is, those males more than four years old who had not acquired the behavior by the age of three were not likely to learn it later. Concerning this last point, KAWAI indicates that by the age of three males move to a peripheral part of the troop and are no longer in close contact with individuals in the central part, which may be called the 'center of diffusion of the behavior'. Such a close social contact is a prerequisite to propagation; and further, it becomes more difficult to acquire a behavior as one gets older. Such general trends are similar to those already pointed out by ITANI [1958] concerning candy diffusion and are the same as the process called 'propagation', instead of 'tradition'.

The second period relates to diffusion observed principally in and after 1959, which differs from that in the first period in the following points and is in no respect different from ITANI's so-called 'tradition'. The sweet potato washing behavior reached its peak in the troop in the 1958–1959 period, and it was no longer a new behavior for the infants born during that period; they learned it, rather, as a normal feeding behavior. They apparently acquired the behavior from their mothers, and a quick and appropriate diffusion took place from elder to younger individuals. This behavior, requiring a skill different from mere eating of candy, was not learned by individuals less than one year of age, but was acquired between the ages of 1–2.5 years by almost all the infants and juveniles. In view of this phenomenon, KAWAI [1964, 1965] regarded a cultural pressure as something which works on younger generations, for which reason the period was designated as 'the period of pre-cultural propagation'.

In connection with the sweet potato washing behavior, AZUMA and YOSHIBA observed an interesting behavior. In the period from 1957 to 1958, monkeys were observed carrying sweet potatoes to the seashore (instead of the fresh water stream), washing them with sea water, and eating them. By the end of 1961, all of those monkeys washing potatoes had acquired a habit of washing potatoes with fresh water *and* sea water. Further, in the later years, it was seen that those individuals who soaked potatoes in sea water were obviously doing this to season the potatoes with salt and not simply to wash off the sand. Such behavior is called 'a seasoning behavior' by KAWAI.

The wheat-washing behavior was called 'a placer-mining behavior' by KAWAI, for the reason that it was the same in principle. The process of diffusion of this behavior into the troop is little different from the sweet potato washing behavior, except that it appeared mostly at the age of two to four years, a little later than was the case in the acquisition of the sweet potato washing behavior. That is, propagation among playmates and propagation along kinship lines appeared conspicuously. Especially in the lineage to which the originator, F-111, belongs, 10 out of 15 individuals, excluding yearlings, performed the behavior. An interesting phenomenon accompanying the wheat-washing behavior was also observed. One monkey would see another washing wheat and would attack him and plunder the wheat that had already been separated from sand in the water and was therefore easier to eat. This 'snatching behavior' was first observed in 1956 by AZUMA, who traced a relation between the gradual increase in the number of individuals performing it and the progress of diffusion of the wheat-washing behavior. According to KAWAI: 'For Japanese monkeys usually do not plunder others' food by attack except for special occasions. Therefore, the snatching behavior has an important implication as a behavior to take advantage of others' labor.'

The bathing behavior and the 'give-me-some behavior' and others will not be described in detail here. As these behaviors spread through the troop, one could see differences in the mode and speed of propagation, depending on the nature and difficulty of the particular behavior itself and the value it has in the life of the monkeys. Figure 4, which is taken from KAWAI's [1965] paper, shows the relation between acquisition of the four behaviors discussed and the age of the monkeys. Concerning this figure, and in particular sweet potato washing and wheat washing, KAWAI [1965] states: 'These behaviors should be oriented as those in the pre-stage of material culture or tool using or a behavior illustrating the process towards these' and, further, 'in the wheat washing behavior a higher intellectual activity is required than the sweet potato washing behavior'. In addition, he points out that the give-me-some behavior

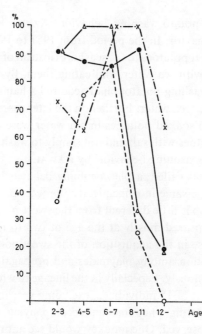

Fig. 4. Correlation between age and acquisition rate of new diffusing behaviors in Koshima troop [after KAWAI, 1965]. ● = Sweet potato washing behavior; ○ = wheat washing behavior; △ = bathing behavior; × = give-me-some behavior.

is an attitude toward men and accordingly the expression of an implicit pre-culture, and that in this respect, as well as in the age of diffusion, it is different from the other three types of behavior.

This detailed study by KAWAI seems to be worthy of special attention for two reasons. (1) It brought nearly to completion the study of culture at Koshima initiated by KAWAMURA and furthered the analysis based on the genealogy of the troop already made clear by KAWAMURA. And (2), it took up the question of intelligence of individual monkeys and the relation of intelligence to group behavior acquisition, and thus made contact between field problems and those of experimental psychology.

Culture as an Intellectual Behavior

A psychological field experiment even more explicitly concerned with this last point was conducted by ATSUO TSUMORI, KAWAI, and RYOJI MOTOY-

OSHI [TSUMORI, 1964, 1966, 1967; TSUMORI, KAWAI and MOTOYOSHI, 1965]. It
is an experiment called 'a sand-digging test'. A hole is dug in the sand before a
monkey; a few pieces of ground nuts are put therein, and the hole is covered
over with sand. The monkey is supposed to dig up the sand and find the nuts –
a kind of simple delayed-response experiment. TSUMORI, KAWAI, and
MOTOYOSHI tried this test because they considered that sweet potato washing
behavior, etc., is related to problem solving. Some important results of this
test are described here.

First of all, TSUMORI reported that although a great difference in acquisition ability was observed between the ages of three and four in the candy test,
a marked decline in acquisition ability was noticed in all groups at ages over
six or seven in the sand-digging test. He pointed out that while the candy test
was a type of conditioned learning, the sand-digging test was a more difficult
type of problem solving for monkeys. In view of the fact that a peak was
reached at the age of six to seven in the sand-digging test, it may be considered
to have a feature in common with the wheat-washing behavior in figure 4. No
sex difference can be observed up to the adolescent stage, and females show
better results at the adult stage. TSUMORI states that though the difference in
intellectual ability between troops cannot be determined on the basis of this
test, in the first trial of the test the Koshima troop attained the rate of 43.9%,
the Takasakiyama troop 31.5%, and the Ohirayama artificial group 23.3%.
These differences suggest that some ecological or sociological factors must be
at work. He further comments that in the same kinship group a trend of rapid
diffusion of the correct solution of this test can be seen, from which it is
obvious that the kinship group serves as a medium for communication of this
type.

The newly acquired behaviors as seen in Koshima have all evolved as
adjustive behaviors, following friendly contacts with men during periods of
provisionization or, in other words, in accordance and sympathy with radical
changes in their environment. Besides such questions as sociological factors,
not yet analyzed fully, and a difference in the manner of contact between
monkeys and men, there is the problem of why only the Koshima troop,
among a number of provisionized troops, has developed a number of conspicuous and 'inventive behaviors', as termed by FRISCH [1963]. Here it is
impossible to disregard the effects of differences in environmental factors;
that is, such environmental elements as the sandy beach and sea present in
Koshima do not exist in the environments in which troops at Takasakiyama
and other areas are living.

The same statement applies to 'the hot-spring-bathing behavior' and 'the apple-washing behavior with snow', as reported by AKIRA SUZUKI [1965] in the Jigokudani troop on the Shiga Heights, a snowy area in Central Japan. The diffusion of behavior in these examples indicates an adjustment of the troop to special environmental elements of the habitat. However, what commands more attention is a food culture quite different from that of monkeys in a warmer area, as reported in similar studies on monkeys in a snowy area by KAZUO WADA [1964] and SUZUKI [1965]. In this area, monkeys pass the winter in a deciduous forest covered deep with snow and for their food depend exclusively on the cambium found in the bark of the branches of deciduous trees.

Culture and Social Structure

In 1970, IMANISHI delivered a lecture on 'L'influence qu'exerce l'environment social du stade immature sur la détermination de la hiérarchie chez les singes Japonais' at the 198th International Colloquium held at Paris. In this lecture he reported on the subsequent conditions of Toku in the Takasakiyama troop. In 1956, at the age of five, Toku finally left the central part of the troop and moved to the periphery. There, among young males, he gradually raised his rank until he had attained the seventh rank among males of the troop in 1965, the fourth rank in 1967, and the second rank in February 1969. Finally in December 1969, when Dandy, who ranked first, disappeared, Toku at the age of 19.5 reached the first rank; that is, he became the most dominant leader. 'Thus my anticipation hit the mark fine', says IMANISHI.

However, too many problems seem to remain to allow us to regard the identification theory as having been verified on the basis of this single instance only. First of all, in 1957 when IMANISHI suggested the identification theory, the social unit of Japanese monkeys was regarded as a closed one. Of course, the existence of solitary males was already known; and instances of males leaving a troop were also on record. But we still continued to accept the scheme that a male born in the central part of a troop moves by the age of three to the peripheral part of the troop and after growing up there, eventually returns to the center and assumes leadership of the troop [ITANI, 1954]. IMANISHI [1957d], too, comments on the fact that among young adult males at Takasakiyama two individuals playing a role of 'preceding and watching' left the troop: 'I cannot understand at all why they made a choice of the way to solitary males where the troop culture, acquired at great pains, seems to be of no more use.'

However, TOSHISADA NISHIDA [1966] subsequently wrote a report that the solitary way is normal for many male Japanese monkeys; and further, in recent summary reports, ITANI [1972] concluded that most male Japanese monkeys left the troop in which they had been born and had grown up. This conclusion does not mean that all male Japanese monkeys waste a troop culture they acquired with much trouble, for it is known that such males, having left the troop, approach another troop after their wandering and become a member thereof [HAZAMA, 1962; WADA, 1964; NISHIDA, 1966; KOYAMA, 1970]. Moreover, examples of males having become leaders in foreign troops are also known [e.g., KANO, 1964]. However, at least the question of 'breeding' as proposed by IMANISHI [1957d], must be corrected by this theory. To clarify this question, one will again have to await the result of long-term, difficult observations to determine whether only an individual who is the son of a dominant female in a troop and who is excellent in leader identification (and leadership identification is hardly an easy thing to measure objectively) can participate in another troop and become a leader there.

As another possibility, one may consider that there are two ways of living for male Japanese monkeys; one is to stay in the troop in which they were born until attaining a high rank, and the other is to leave the troop while they are young [ITANI, 1972]. The former way is probably followed by less than 10% of male monkeys. In the meantime, why did only Toku among so many males in Takasakiyama remain in the troop? There were a number of dauntless males with a violent temper like Jupiter, the former leader, and there were some monkeys possessing a serene dignity like Titan. All of these males left the troop while they were young. The question entertained by ITANI as to why vigorous, splendid young males, playing progressively different social roles, first left the troop also served as a motive for an intensive study of solitary males by NISHIDA [1966]. Though it is merely an intuitive impression gained in an attempt to find out the social significance of only 10% of the males staying within a troop, it seems that Toku has few characteristic features compared with former leaders in the Takasakiyama troop, being rather inferior to them. The same seems to apply to Nula, currently ranking seventh, the only one besides Toku staying in the troop of the 44 males that had been identified individually by ITANI in 1955. At least it may be said that those males leaving the troop of their own accord possessed more masculine elements.

The study of Japanese monkeys, starting from the hypothesis of culture by IMANISHI [1952], has been continued over 25 years up to the present. Needless to say, not all young students have followed his theory. However, it is certain that the profound questions and theories proposed by IMANISHI have always stimulated the study of Japanese monkeys. The culture theory has been one of the moving forces behind this work. In continuing the present

study, the authors recognize that IMANISHI's theory still has a modern significance and presents a number of fresh questions. In the latter half of the 1960's,
the question of culture ceased to be of great concern. Many researchers have
come to take up instead studies of basic social structure and behavior. However, the results thereof will undoubtedly be collated with various questions
mentioned in this paper and reviewed from a new viewpoint. On the other
hand, the authors have not yet completed the work of elaborating in detail and
along the time axis the bulky data on monkey life-history collected over a
period of these last 20 years. Further, the recent conclusion that the social
unit of a troop is not a closed one requires us to re-examine all the results obtained up to the present; and this requirement includes the results pertaining
to culture.

References[6]

FREUD, S.: New introductory lectures on psychoanalysis (Norton, New York 1933).
FRISCH, J.: Research on primate behavior in Japan. Amer. Anthrop. *61*: 584–596 (1959).
FRISCH, J.: Japan's contribution to modern anthropology; in ROGGENDORF Studies in
     Japanese culture, pp. 225–244 (Sophia University, Tokyo 1963).
HAZAMA, N.: On the weight-measurement of wild Japanese monkeys at Arashiyama. Bull.
     Iwatayama Monkey Park No. 1 (1962)* [Partial Eng. transl.: Weighing wild Japanese
     monkeys in Arashiyama. Primates *5*: 81–104, 1964].
IMANISHI, K.: Evolution of humanity; in IMANISHI Man (Mainichi-Shinbunsha, Tokyo
     1952)*.
IMANISHI, K.: Problems in the study of Japanese monkeys. Shizen *12 (2)*: 3–9 (1957a).*
IMANISHI, K.: Social behavior in Japanese monkeys, *Macaca fuscata*. Psychologia *1*: 47–54
     (1957b).
IMANISHI, K.: Learned behavior of Japanese monkeys. Jap. J. Ethnol. *21*: 185–189 (1957c).*
IMANISHI, K.: Identification – a process of socialization in the subhuman society of *Macaca
     fuscata*. Primates *1*: 1–29 (1957d)* [Eng. transl.; in IMANISHI and ALTMANN Japanese
     monkeys, a collection of translations, pp. 30–51 (University of Alberta Press, Edmonton
     1965)].
IMANISHI, K.: Social organization of subhuman primates in their natural habitat. Current
     Anthrop. *1*: 393–407 (1960a).
IMANISHI, K.: Bird, monkey and man; is it possible to build a general theory to support
     'identification'? Jinbun Gakuho *10*: 1–24 (1960b).*
IMANISHI, K.: L'influence qu'exerce l'environnement social du stade immature sur la déter-
     mination de la hiérarchie chez les singes Japonais. Colloques internationaux du
     CNRS, No. 198, pp. 149–154 (Paris, 1971) [Jap. transl.; in IMANISHI Animal society,
     pp. 354–360 (Shisakusha, Tokyo 1972)].

6   Works marked with asterisk (*) are published in Japanese.

ITANI, J.: Japanese monkeys in Takasakiyama; in IMANISHI Nihon dobutsuki, vol. 2 (Kobunsha, Tokyo 1954).*

ITANI, J.: The prehuman language. Shizen 11 (11): 22–27 (1956a).*

ITANI, J.: Food habits of the Japanese monkeys. I. Vegetable food (Primates Research Group Private Press, Kyoto University, Kyoto 1956b).*

ITANI, J.: Personality of Japanese monkeys. Iden 11 (1): 29–33 (1957).*

ITANI, J.: On the acquisition and propagation of a new food habit in the troop of Japanese monkeys at Takasakiyama. Primates 1: 84–98 (1958)* [Eng. transl.; in IMANISHI and ALTMANN Japanese monkeys, a collection of translations, pp. 52–65, University of Alberta Press, Edmonton 1965].

ITANI, J.: Paternal care in the wild Japanese monkeys, Macaca fuscata fuscata. Primates 2: 61–93 (1959).

ITANI, J.: Vocal communication of the wild Japanese monkey. Primates 4: 11–66 (1963).

ITANI, J.: Social structure of primates (Kyoritsu Shuppan, Tokyo 1972).*

ITANI, J. and HAZAMA, N.: Development and mother-baby relationship of Japanese monkeys. Physiol. Ecol. 5: 42–51 (1953).*

ITANI, J. and TOKUDA, K.: Monkeys on Koshima Islet; in IMANISHI Nihon dobutsuki, vol. 3 (Kobunsha, Tokyo 1958).*

KANO, K.: On the second division of the natural troop of Japanese monkeys in Takasakiyama; in ITANI et al. Wild Japanese monkeys in Takasakiyama (Keiso Shobo, Tokyo 1964).*

KAWABE, S.: Monkey babies (Chuokoronsha, Tokyo 1964).*

KAWABE, S.: A study of Japanese monkeys raised in isolation: development of Quol and Rika for the first 8 months after their birth; in KAWAMURA and ITANI Monkeys and apes – sociological studies, pp. 403–449 (Chuokoronsha, Tokyo 1965).*

KAWAI, M.: Ecology of Japanese monkeys (Kawade Shobo, Tokyo 1964).*

KAWAI, M.: Newly acquired pre-cultural behavior of a natural troop of Japanese monkeys on Koshima Island. Primates 6: 1–30 (1965).

KAWAMURA, S.: On a new type of feeding habit which developed in a group of wild Japanese macaques. Seibutsu Shinka 2 (1): (1954).*

KAWAMURA, S.: Prehuman culture, Shizen 11 (11): 28–34 (1956).*

KAWAMURA, S.: The process of sub-culture propagation among Japanese macaques. Primates 2: 43–60 (1959).

KAWAMURA, S.: Sub-culture among Japanese macaques; in KAWAMURA and ITANI Monkeys and apes – sociological studies (Chuokoronsha, Tokyo 1965).*

KLUCKHOHN, C. and KELLY, W. H.: The concept of culture; in LINTON The science of man in the world crisis (Columbia University Press, New York 1945).

KOYAMA, N.: On dominance rank and kinship of a wild Japanese monkey troop in Arashiyama. Primates 8: 189–216 (1967).

KOYAMA, N.: Changes in dominance rank and division of a wild Japanese monkey troop in Arashiyama. Primates 11: 335–390 (1970).

LINTON, R.: The cultural background of personality (Appleton Century Crofts, New York 1945).

LORENZ, K.: King Solomon's ring (Borotha-Verlag, Vienna 1952).

MAEKAWA, S.: On the study of an artificial group consisting of isolated and man-reared

Japanese monkeys; food habits in the new environment, Tsurushima Island. Nankaiseibutsu *9 (2):* 48–53 (1967).*

MILLER, N. E. and DOLLARD, J.: Social learning and imitation (Yale University Press, New Haven 1941).

MIYADI, D.: On some new habits and their propagation in Japanese monkey groups. Proc. 15th int. Congr. Zool. London, pp. 857–860 (1959).

MIZUHARA, H.: Japanese monkeys (Sanichi Shobo, Kyoto 1957).*

MOWRER, O. H.: Learning theory and personality dynamics (Ronald, New York 1950).

NISHIDA, T.: A sociological study of solitary male monkeys. Primates *7:* 141–204 (1966).

SUZUKI, A.: An ecological study of wild Japanese monkeys in snowy areas – focused on their food habits. Primates *6:* 31–72 (1965).

TSUMORI, A.: An analysis of the intellectual behavior of wild Japanese monkeys at Takasakiyama by the sand-digging test; in ITANI *et al.* Wild Japanese monkeys in Takasakiyama, pp. 74–90 (Keiso Shobo, Tokyo 1964).*

TSUMORI, A.: Delayed response of wild Japanese monkeys by the sand-digging method. II. Cases of the Takasakiyama troops and the Ohirayama troop. Primates *7:* 363–380 (1966).

TSUMORI, A.: Newly acquired behavior and social interactions of Japanese monkeys; in ALTMANN Social communication among primates (The University of Chicago Press, Chicago 1967).

TSUMORI, A., KAWAI, M., and MOTOYOSHI, R.: Delayed response of wild Japanese monkeys by the sand-digging test. I. Case of the Koshima troop. Primates *6:* 195–212 (1965).

WADA, K.: Some observations on the life of the Japanese monkeys in the snowy district in Japan. Physiol. Ecol. *12 (1–2):* 151–174 (1964).*

YAMADA, M. (KAWABE): A case of acculturation in the subhuman society of Japanese monkeys. Primates *1:* 30–46 (1957).*

YAMADA, M. (KAWABE): Five natural troops of Japanese monkeys in Shodoshima Island. I. Distribution and social organization. Primates *7:* 315–362 (1966).

YERKES, R. M.: Chimpanzees: a laboratory colony (Yale University Press, New Haven 1943).

Authors' addresses: Dr. JUNICHIRO ITANI, Laboratory of Physical Anthropology, Faculty of Science, Kyoto University, *Kyoto* (Japan); Dr. AKISATO NISHIMURA. Primate Research Institute, Kyoto University, *Inuyama, Aichi 484* (Japan)

Symp. IVth Int. Congr. Primat., vol. 1: Precultural Primate Behavior,
pp. 51–75 (Karger, Basel 1973)

Testing for Group Specific Communication Patterns in Japanese Macaques

G. R. STEPHENSON

Department of Zoology, University of Wisconsin, Madison

Introduction

A primary feature of primate sociobiology emergent from recent studies
of cercopithicine monkeys and apes is the wide variation in behavior by means
of which individuals of each particular social group within a species adapt to
their respective environments [see JAY, 1968]. Variations in behavior do not
appear simply as isolated instances, but appear as clusters of particular
patterns (as recurrent styles or types) of behavior specific to particular social
groups. Examples of group-specific patterns of behavior include acceptance
of or preference for specific food types, special behavior to procure particular
food items, and avoidance of particular objects and creatures in the environment [see HALL, 1963a, and STEPHENSON, in press, for many examples]. Routes
of movement through the environment also appear to be group-specific
[KOFORD, 1963; STEPHENSON, 1968]. In addition, some aspects of social
structure and relative rank relations in terms of age/sex classes appear to be
specific to particular groups [KAWAMURA, 1958; HALL und GOSWELL, 1964].

The observed variability in patterns of adaptive behavior across social
groups within a species has been examined in terms of variation in the physical
components of the environment [HALL, 1963b; GARTLAN and BRAIN, 1968;
KUMMER, 1968; CROOK, 1970a] and in terms of variation in social traditions
[KAWAMURA, 1959; KAWAI, 1965; FRISCH, 1968; HALL, 1968; STEPHENSON,
1968, in press; CROOK, 1970b]. Sociobiological mechanisms underlying the
maintenance of social traditions have been discussed in terms of observational
learning [HALL, 1963a; HALL and GOSWELL, 1964] and admonition [STEPHEN-
SON, 1967]. Adaptation through social traditions assumes that (a) individuals
of the species have the capacity to learn, (b) individuals live in social groups,

and (c) the behavior of individuals can be shaped in the course of social inter-action with members of their social group. Given these prerequisites, mani-festation of the capacity for social tradition or culture appears to an observer as a set of group-specific behavior patterns.

For a species which inhabits a variety of environments, the capacity for maintenance of social traditions would facilitate the modification of certain relevant behavior specifically to accommodate local conditions without loss in plasticity of the underlying neural and morphological bases of the accom-modation. A model of adaptation among nonhuman primates through social traditions has been presented elsewhere [STEPHENSON, in press]. In this model, the exploratory nature and ready learning capacity of young members of a primate social group are seen as requisite complements to the behaviorally conservative nature of older members, such that the set of cultural patterns is shaped and maintained with optimal relevance to the environment of indi-viduals in the group. Culture as a set of learned behavior and products of learned behavior patterns [e.g., termite sticks prepared by wild chimpanzees; SUZUKI, 1966], manifested in accordance with a set of acquired perceptual rules [in the sense of learning sets *à la* HARLOW, 1949, and transformations *à la* LENNEBERG, 1967], appears to serve as the primary adaptive mechanism of several species of primates in addition to humans.

For a species of primate in which culture has emerged as the primary adaptive mechanism, maintenance of the social group *per se* gains further biological significance [see ETKIN, 1964, pp. 1–34, for a general overview of the effects of living in groups]. In contradistinction to the notion of culture as 'superorganic', it is now recognized that individuals of a social group main-tain and perpetuate cultural behavior via communication between and among their fellows [e.g., OPLER's view of human beings as the vehicle of social tradition; OPLER, 1964]. Next to human beings, macaque species have been the most intensively studied primates. As among human beings, communi-cation patterns (i.e., recurrent stylized behavior setting up communication signals) serve to mediate interactions among individuals [CHANCE and MEAD, 1953; ALTMANN, 1965] and thereby facilitate the cohesiveness requisite to maintenance of the social group [IMANISHI, 1960]. Indeed, ALTMANN [1962, p. 279] delimits a social group as '...bounded by frontiers of far less frequent communication'. It was in light of the apparent importance of communication patterns in maintenance of the social group, which is the prime prerequisite to adaptation through culture, that the present study of group-specific com-munication patterns in Japanese macaques *(Macaca fuscata)* was undertaken in 1968.

Of particular interest here are the degree and distribution of variability in the form and function of communication patterns across social groups of Japanese macaques. Variability in form and function and in the relation between them appears to range along a phylogenetic-ontogenetic continuum [MARLER, 1961; SKINNER, 1966] of increasing openness [SMITH, cited by SEBEOK, 1965] from the uniformity and ubiquity of highly stereotyped and speciesspecific patterns [TINBERGEN, 1951] to the diversity and group-specific distribution of arbitrary patterns [arbitrary in the semantic sense; HOCKETT, 1958, p. 354]. At one end of the continuum is the threat facial expression of an adult rhesus macaque, which, when projected as a photograph, induces fear responses in naive isolate-reared infants [SACKETT, 1966]. The stochastics of patterns of behavior produced by electrical stimulation of brain centers in female rhesus inducing mounting responses in unwired males [DELGADO, 1965, p. 32] similarly suggest species-specific relations between form and function of some communication patterns. These kinds of patterns imply contemporaneous selection during evolution of both pattern and response, that is, a parallel molding of both the behavior of the sender [perhaps via ritualization; see TINBERGEN, 1952, on derived activities and HUXLEY, 1966, on ritualization] and the information processing network in the central nervous system of the receiver [HUBEL, 1963; HUBEL and WIESEL, 1963; in the sense of LETTVIN et al., 1959, and KONORSKI, 1967]. On the other end of the continuum is the unlikely 'backflip' pattern used by an informant monkey to transmit information on the solution of a novel problem to a naive operator in a visually accessible nearby cage [MASON and HOLLIS, 1962]. This kind of pattern appears to arise by convention (i. e., become established through usage) and is learned by each of the individuals that perform it or that respond appropriately to its performance as a communication pattern. The 'acquaintance behavior' of irus macaques [THOMPSON, 1967], the 'affectional present' of rhesus macaques [HANSEN, 1962], and patterns used in the communication of affect in rhesus macaques [MILLER et al., 1959, 1966; MILLER, 1967] presumably lie somewhere between these extrems. Openness, hence arbitrariness of form and function and the relation between them, might have been expected to predominate in primates as compared to lower vertebrates [SEBEOK, 1965, p. 1010]; but before the present study, useful details were not available.

My warmest thanks go to Prof. JOHN T. EMLEN for discussion and guidance in all phases of the study; to Dr. MUNEMI KAWABE of Osaka City University and his students, Mr. KOSHI NORIKOSHI, Mr. AKIO ORII, and Mr. SADASHIGE ASHIZAWA, for collaborating in the field work; to Mr. JACK LUND of the University of Wisconsin Laboratory of Photography for his beautiful cinematography; and to my wife, GAYLE, for assistance in gathering the data and for patience during their analysis. Thanks also to Mr. S. IWATA, the Miyajima Ropeway Company, and the Japan Monkey Center for their cooperation at the study sites; to Mr. Y. KUNIZAWA, Mr. S. HYOKA, and Mrs. S. MITO, for their help at the sites; to Mr. STEVEN GREEN of Rockefeller University and his assistant, Miss KAREN MINKOWSKI, in the field work; and to Miss MARY KAY DINEEN, Miss DEBORAH YOSHIHARA, and the 'fourth floor group' in the Department of Zoology, University of Wisconsin, especially Mr. TIMOTHY JOHNSTON and Mr. CRAIG MACFARLAND, for their assistance and advice in analysis and presentation of the data. This study was supported by the National Science Foundation (grant GB-7551), The Japan Society for the Promotion of Science, The Wilkie Brothers Foundation of Des Plaines, Ill., The New York Zoological Society, and the Graduate School of the University of Wisconsin.

Monitoring Communication Events

The data for this study are instances of patterned vocal and nonvocal behavior monitored by human observers as they occurred in the course of social interaction among Japanese macaques. Patterned nonvocal behavior, as used here, refers to stylized motor acts which serve to set up signals in the visual modality in recurrent signal forms (as perceived types or kinds) [ZUSNE, 1970]. Such behavior includes postures, gestures, facial expressions, oriented positionings, and changes in interindividual spatial relations. Instances of patterned behavior in social settings are regarded as communication events in a system of communication channels between individual macaques.

In terms of telecommunication theory, a communication channel has three primary components: a sender, a signal, and a receiver [CHERRY, 1966]. A signal is a physical disturbance of the medium between sender and receiver, such as the sound wave pattern set up in the course of vocal behavior or the pattern of reflected light set up in the course of nonvocal behavior. A signal is set up by a sender, here the performer of the vocalization or facial expression serving as source of the signal. A receiver is an individual that responds to the vocalization or facial expression or, more generally, is physiologically and situationally capable of sensing a signal set up by a sender.

In the field situation, the human observer of a communication event among nonhuman animals generally locates and monitors the components of the communication channel by keying on the signal. If the signal is in the visual modality, the sender is typically coincident in space and time with the apparent source of the signal [MARLER, 1965]; and in practice, the apparent source can be reliably designated as the sender. When the signal is in the auditory modality and several potential senders are present, the inherent difficulties of localizing the source of the sound often limit the observer to guessing which individual is setting up the signal. Designating the receiver is even more problematic. There can be many receivers for any one signal set up by a sender. For any communication event monitored in the field situation the receiver can only be inferred, although the inference may be strong when an experienced observer sees an individual other than the sender behave in an expected manner subsequent to transmission of a signal. Within these limitations, the field methods of the present study focused on the signals and the senders, receivers, and situations of communication events among individuals in three free-ranging troops of Japanese macaques.

A Semiotical Consideration of the Communication Event

To elucidate socially acquired communication patterns in the repertoire of a species in which members live in social groups and have a demonstrated capacity for social acquisition of arbitrary signs, it is necessary at all times in the course of study to take account of an inherent duality in the relation between a pattern and the context in which it appears. In this study, semiotic, the general theory of signs, provided the philosophical foundations for developing and implementing controls in the field procedures and laboratory analyses of the visual communication signals recorded on film. A semiotical consideration of the communication event has been presented in detail elsewhere [STEPHENSON, 1973] and is merely outlined here.

The central notion of semiotic is the sign. Briefly, the sign is 'the something which stands to somebody for something in some respect or capacity' [CHARLES S. PEIRCE, quoted in HARTSHORNE and WEISS, 1931–1935, vol.2, § 228]. That which the sign stands for is called the designatum, and that which the sign effects or expresses in an interpreter is called the interpretant [from GOUDGE, 1969, after PEIRCE][1]. The relation between these three axiomatic elements of semiotic has been viewed as an inseparable triad in which the elements are defined in terms of one another. Any particular 'something' can serve as any one of the elements in the triad, its particular service being determined solely by its relation to the other two elements in the triad.

An analytic approach to the communication event in terms of this philosophical system was initially developed by CHARLES MORRIS [1938]. MORRIS' later works [e.g., 1946 and 1964] include assumptions that in general limit

1   See T. A. GOUDGE, 1969. [Ed.]

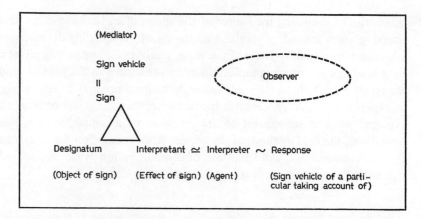

Fig. 1. The observer's relation to the semiotic triad. The axiomatic elements of the triad are not directly observable, but are monitored through the sign vehicle and the response of an interpreter. The response sets up a signal that serves as a sign vehicle of the effect or of a particular taking account of the sign in the interpreter.

their application to human beings studying human beings [cf. WYKOFF, 1970, p. 60; but see also MARLER, 1961, p. 300]. By treating the sign as a sign vehicle and allowing the monitorable response of an interpreter to stand for the interpretant, MORRIS was able to define a position for the observer from which he can begin to assess a communication event as a mediated taking account of 'the something which stands to somebody for something in some respect or capacity' (fig. 1).

As SMITH [1968, p. 46] has pointed out, to analyze a communication event in terms of semiotic, it is necessary to consider it, not as one, but as many instances of semiosis (the process in which something functions as a sign). A signal, as a physical disturbance of the medium, may be set up by a sender in the course of a behavior that stands for something to the sender. From another point of view, a receiver may take a signal as standing for something. A relation between a sender's behavior and the effect of its behavior on a receiver can be carried by the signal serving as a sign vehicle. A sign vehicle is the physically monitorable aspect of a sign. It serves to carry a semiotic relation between sender and signal and signal and receiver which resides in the similarity of the respective alphabets of signs within sender and receiver [cf. CHERRY, 1966, p. 171]. The capacity for carrying this relation rests on the fundamental property of semiotic processes, namely, their infinite regressiveness [WYKOFF, 1970]. Infinite regressiveness stems from the interchangeability of the sign, designatum, and interpretant, whereby the interpretant of any

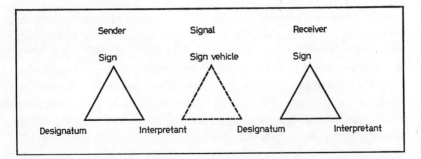

Fig. 2. The infinite regressiveness of semiosis. Any sign can serve as designatum or interpretant of another sign; hence any two signs may be connected by an intervening and indeterminately long series of related signs. Although the initial perception of a signal may be within-brain and unassessable by an observer, the apparently ultimate interpretant may be a signal assessable by the observer as well as a conspecific receiver. The apparently ultimate interpretant in the receiver may be a behavior also monitorable by the observer through the signal it sets up. From the observer's point of view, the signal would appear to serve as a sign vehicle in that it carried a relation between the designatum of the sender and the interpretant of the receiver.

sign can serve in other semioses as the sign or as the designatum of another sign, the sign can serve as designatum or interpretent, and the designatum as the sign or interpretant. With respect to the communication event, the behavior of a sender may serve as interpretant for the sender (as the effect of processing a sign) and, by means of the sign vehicle, may serve as designatum for the receiver (fig. 2).

As a mere monitor of communication events in another species, the human observer is very limited in what he can infer about the semiotic properties of a sign vehicle residing in a signal set up by a sender and apparently perceived by a conspecific receiver. Most of the processing of signs is within-brain and not directly monitorable by the observer. The monitorable aspects of a communication event (as a set of instances of semiosis) are the signal in which the sign vehicle resides and the context in which the signal is set up.

The context of a signal has been viewed as being the combined influence of the immediate context and the historical context on the receiver's perception of the signal [Smith, 1965]. Immediate context refers to the total sensory input to the central nervous system of the receiver at the time the particular signal is sensed. Historical context refers to phylogenetic prepotencies and the past experience of the receiver which modify his perception of the signal. From the receiver's point of view, the notion of context '... refers to anything which can be thought of as accompanying a signal' [Smith, 1965, p. 405]. From the human observer's position outside of the channel of communication he is monitoring [CHERRY, 1966, fig. 3.2, p. 92], data pertaining to the historical context are largely unassessable in the field situation. The aspect of context which is assessable by the observer is the immediate context within which the observer as another receiver senses the signal set up by the sender. The observer's perception of the signal is also modified by an historical context residing in himself and based upon his past experience with the species and with the particular individuals involved in the communication event. In the present study, context refers to the particular individuals participating as sender and receiver in the communication event; their physiological (seasonal) condition; their sex and age; their genealogical relations; their dominance-subordination relations in light of dependent rank; objects in the environment of the event to which the signal may have referred; location of the event with respect to daily movement; time of the event with respect to feeding schedules; daily routine, and preceding (and, in retrospect in the course of analysis, succeeding) behavior; and any other information qualifying the environment of the event.

An observer must rely on analysis of the relations between signals and contexts as they appear to him in order to discern the sign vehicles and develop an alphabet of signs or a code applicable to communication events of the species from his viewpoint outside of the channel. With experience, an observer may come to know and key on the sign vehicle and/or to know and key on a particular subset of contextual data which appears to correlate by rules of usage [MORRIS, 1938, p. 48] with the occurrences of a particular sign vehicle. Enquiry may begin on the pragmatic dimension of semiotic, which deals with the relations of signs to their users, or more generally, includes '... all the psychological, biological, and sociological phenomena which occur in the functioning of signs' [MORRIS, 1938, p. 30]. Abstracting from the pragmatic dimension, enquiry may proceed to examine the relation of signs to their designata [CHERRY, 1966, pp. 221–227] on the semantic dimension and the 'formal relations of signs to one another' [MORRIS, 1938, p. 6] on the syntactic dimension. Of course, as MORRIS pointed out [1938, p. 33], 'In a systematic presentation of semiotic, pragmatics presupposes both syntactics and semantics, as the latter in turn presupposes the former, for to discuss adequately the relation of signs to their interpreters requires knowledge of the relation of signs to one another and to those things to which they refer their interpreters.'

The relations of signs to their designata are of particular concern in the present study because macaques have demonstrated the capacity for social acquisition of arbitrary signs [STEPHENSON, in press]. Arbitrariness appears on the semantic dimension whenever the relation between the form of a sign vehicle and its apparent designatum is 'open', that is, unexpected or not implied in the form *per se*. The relation between an arbitrary sign and its designatum is not carried by a physical or geometrical resemblance between the form of the sign vehicle and the form of the something for which the sign stands [HOCKETT and ALTMANN, 1968, p. 63], but relies wholly on the legisign, as manifest in the rules of usage, to carry the relation [cf. GOUDGE, 1969, p. 146, on PEIRCE's notion of symbol]. This is in contradistinction to the relation between an iconic sign and its designatum, where the form of the sign vehicle involves its rule of usage and thereby implies the designatum. Arbitrary signs are also distinguished from indexical signs such as pointing, where the incidence of the sign vehicle is 'connected' (in space and time) to the designatum of the sign. Indexical and iconic signs are more readily interpretable than are arbitrary signs and are, therefore, more likely to come under selective processes such as ritualization [*à la* TINBERGEN, 1952]. For arbitrary signs, the particular relation a sign vehicle carries is most likely to be a function of the animal's individual experience [i.e., the experiential component of the previously mentioned notion of historical context; SMITH, 1965]. Given that the experience of an individual living in a social group can be structured in the course of social interaction with other members of the group (i.e., culturally), particular relations between signals and contexts could arise through usage and become established as particular relations between sign vehicles and the things for which the signs stand. The 'backflip' of an informant monkey in the MASON and HOLLIS [1962] study of communication among young rhesus macaques provides a clear laboratory example of a socially acquired arbitrary sign. Arbitrary signs would appear to arise most readily by convention, i.e., through usage, and to propagate through the membership of a social group through social learning. Elucidation of arbitrary sign vehicles (on the semantic dimension of semiotic) would, therefore, provide the most positive demonstration (on the pragmatic dimension) of a cultural basis for group-specific communication patterns among nonhuman primates.

For the observer of communication events of a species with the capacity to form arbitrary signs, the clue as to whether a signal with apparently arbitrary form is serving as a vehicle of a socially acquired sign lies in the duality of the signal-context relation. To take account of the potential duality of any signal-context relation, the observer's observations and analyses of communication events among members of such a species must take account of the signals and the contexts as if they were varying independently of one another. Given that communication among members of a social group is optimized by propagation of the rules of usage of arbitrary signs throughout the membership of the group, there are four logical possibilities for the relation between signal form and context when manifestations of two socially acquired arbitrary sign vehicles are observed in more than one social group of such a species. The possibilities are: (1) two signals similar (that is to say, if different, indistinguishably so), two contexts similar; (2) two signals different, two contexts different; (3) two signals different, two contexts similar; and (4) two signals similar, two contexts different. For observations fitting the first two possibilities, the observer cannot address himself to the question of social acquisition without further data (e. g., diachronic data on the origin and paths of propagation of a particular form of sign vehicle within the social group) [STEPHENSON, in press]. In observations fitting the third and fourth possibilities, given evidence that the social groups share a common gene pool (or, more generally, that they are indeed within the same species), the observer would have sufficient data to suggest that the signs relating the apparently arbitrary sign vehicles and their designata were socially acquired.

The potential duality of the signal-context relation can be taken into account by sampling signals that occur in a particular or known context or by sampling the contexts in which a particular signal form occurs. Since cine matography is a technique which provides highly reliable intersubjective records of behavior, I chose to begin this investigation of communication among free-living macaques by focusing on the signals and concentrating initial efforts on recording, analyzing, and comparing signal forms across social groups. Efforts were, therefore, made to select a control context within which communication events were recorded. To this end, given historical integrity as a social group, the primary criterion in the selection of study troops was the extent of background information on troop sociology available to us from the Japanese primatologists. This information provided control over the social aspects of any context we might wish to sample. A secondary consideration was the conditions for synchronized-sound cinematography at the troop sites.

Control of Context

To optimize the chances of elucidating group-specific communication patterns in Japanese macaques, it is desirable to choose a control context wherein sign vehicles occur with forms most likely to have been delineated through social experience and thereby most likely to vary in form across troops. On the other hand, given the potential duality of signal-context relations in this species, it is necessary at the same time that the control context be defined in terms which are equally applicable across the potentially different traditions of each troop. The courtship context appeared to be the context which best met both of these criteria. Studies of the effects of social isolation on sexual behavior and breeding success among laboratory rhesus macaques [SENKO, 1966; MITCHELL, 1968] suggested that early experience is a primary variable in delineating the ultimate forms of typical pre-copulatory behavior in a species congeneric with the Japanese macaque. At the same time, the courtship context can be defined in terms of the consort relation, a naturally occurring and readily discernible relation across different troops. The consort relation is defined as the relation between a male and a female when they are closely and persistently associated and engaging in serial mounting with regular intermount intervals in a fashion which typically culminates with ejaculation. The courtship context is defined as the setting in which the behavior of a male and a female interacting with one another leads to a decrease and stabilization in interindividual distance, a situation that in turn leads to the formation of a consort relation between the male and the female. (In the consort relation, interindividual distance holds at nearly zero for relatively long periods.)

With the guidance of our Japanese colleagues, the three study sites shown in figure 3, Miyajima, Arashiyama, and Koshima, were so selected that the peaks of mating activity in the associated provisioned troops were offset one from the next by roughly five weeks. By moving from one site to another in succession, we could record behavior of individuals that were in relatively similar physiological (seasonal) condition. To determine whether we were indeed there at the peaks, conception dates were estimated by subtracting 175 days from the observed date of parturition in the subsequent birth season for the breeding females in each troop. [KAWAI has estimated the length of gestation in Japanese macaques at 170–180 days; cf. NAPIER and NAPIER, 1967, p. 406, and KAWAI, 1966.] About 38% of the conceptions at Miyajima probably occurred during the study period there, 62% at Arashiyama, and 40% at Koshima. As distinctive behavioral indices of the courtship context became apparent during the study period at each troop, they were employed to guide the sampling and recording of behavior that appeared to serve in setting up signals as sign vehicles for social communication.

Due to the large variety of visual sign vehicles, the apparently wide variation in signal form, and the contextual complexities stemming from the sophisticated social structure of Japanese macaques, there appeared to be

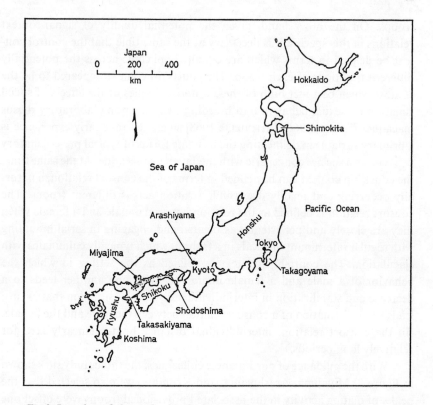

Fig.3. Location of the study sites in Japan. Miyajima is near Hiroshima in south western Honshu; Arashiyama is near Kyoto; and Koshima is off the south-eastern coast of Kyushu.

neither enough time nor enough film to sample the courtship behavior of all individuals in each troop. I concentrated, therefore, on individuals more than four years old and considered to be in the socio-spatial central part of the troop [IMANISHI, 1960]. 'Central' here includes all females and the leader and subleader males in the troops and excludes the peripheral and solitary males associated with the troops. These restrictions were based upon the assumptions that (a) before sexual maturity (in females at 3½ years and in males at 4½ years), individuals are less likely to engage in typical courtship than they are afterward; and (b) there is greater homogeneity in the production of sign vehicles by individuals that have had the opportunity and the time to shape one another's behavior through close association in the central part of the troop. In addition, individuals of the central part of the troop are more likely

Table I. Numbers of central males and females available and of males, females, and discrete pairs observed in study populations

| | A. Available | | B. Observed | | |
	males	females	males	females	pairs
Miyajima	6	27	6	18	49
Arashiyama	9	41	9	29	87
Koshima	6	21	6	17	58

to be located and observed on a regular basis than are other individuals in the troop. The numbers of these individuals available for observation at each troop are presented in table IA. The proportions of males and females are similar across troops ($\chi^2 = 0.228$; d.f. = 2; p >0.50). The proportions of males, females and discrete pairs actually observed (table IB) during the study are also similar across troops ($\chi^2 = 0.409$; d.f. = 4; p >0.80). The behavior of any other animals interacting with the central animals was also recorded on film and in the notes.

Signals were recorded on Fuji black and white negative 16-mm film (type RP) with an Arriflex (model M) cine-camera mounted on a Ceco hydrofluid head tripod with an Angineaux 12–120-mm zoom lens and 400-ft magazine. Film segments were exposed in the course of nearly continuous observations entered by voice on magnetic type during the observation sessions. Nearly 83,000 ft of film were exposed in the mating season: in 132 h of watching monkeys over 26 days at Miyajima, 9.25 h of film were exposed in 293 segments in the course of 48 h of intensive and detailed observations recorded on tape; in 185 h over 48 days at Arashiyama, 12.78 h of film were exposed in 337 segments in the course of 63 h of tape-recorded observations; and in 115 h over 33 days at Koshima, 16.29 h of film were exposed in 446 segments in the course of 79 h of taped observations. About 20% of the film was exposed while two cameras were operating simultaneously from complementary perspectives. In a parallel study, Mr. STEVEN GREEN of Rockefeller University recorded vocalizations. About 30% of the film was taken with synchronized sound recordings. Our Japanese collaborators, Mr. A. ORII at Miyajima, Mr. K. NORIKOSHI at Arashiyama, and Mr. S. ASHIZAWA at Koshima, worked with us at their respective sites and provided ecological, historical, and sociological background and insight into the on going behavior that were fundamental to the refinement of context required for this study.

Sampling was directed toward recording on film as great a variety and as many examples of sign vehicles as could be found in the courtship context. Since ascertaining the context of the communication event was more critical in this study than was estimating the frequencies of signal forms, we followed individuals for long periods in order to accurately evaluate the setting and ongoing relationship between senders and receivers rather than sampling the behavior of individuals in randomized blocks over uniform time intervals. To compensate and maintain a better overview of activity of the troops, observations of courtship between individuals interacting in the vicinity of those on which we were concentrating at the moment were also entered into the record. The quality of these latter observations is generally lower than those of animals on which we were focusing our cameras and attention. Usually, we followed an individual or an interacting pair through the courtship period and between mount series after a consort relation had become established. To elucidate individual contributions to the observed variations in forms of sign vehicles, we followed subjects as they changed consort partners, sampling the behavior of particular females over the weeks of the study period as they courted and consorted with a series of different males and, vice versa, sampling the behavior of particular males as they courted and consorted with one female after another.

Upon return from the field in October, 1969, signals on film were initially studied in terms of signal forms *per se*, without specific reference to the transcribed notes. All of the film was reviewed on a Lafayette Analyst projector at normal (24 frames/sec), faster, and slower rates of projection, both forward and backward, down to one frame at a time. After initial provisional categories of general signal forms were determined from intensive review of the Arashiyama films, the categories were applied to the discernible sign vehicles in the Miyajima films. The categories were adjusted as necessary as new forms were found in the Miyajima films. Adjustments were made again when the provisional categories were applied to the Koshima films. The Arashiyama films were catalogued last, after the categories were consistent and general enough to be applied across the three troops. The films from each troop were then reviewed and described in terms of the revised general categories in conjunction with the corresponding field notes, lineage relationships, social structure, and special social relationships of senders and receivers in the communication events in which signals serving as sign vehicles were recorded.

To sort for signals recorded in the courtship context, sexually receptive periods were delineated for females as blocks of days during which they showed evidence of having participated in serial mounting. In the Japanese macaque, mounting is a cooperative complex of behavior; a female can refuse to be mounted by not raising her hind quarters into the sexual present posture, thereby preventing penile intromission and thrusting. Criteria for delineating the receptive periods were an observed series of four or more mounts with thrusting and regular intermount intervals and an observed ejaculatory mount or the presence of a semen plug, or both. All interactions between males and females recorded during the females' receptive periods were initially classified as courtship encounters. During review of the data, the general sequence of the behaviorally classified types of courtship encounters recurring with respect to mount series or ejaculations suggested refinement of the courtship context into four stages. The stages are: (1) advertising and monitoring availability for mating; (2) testing and closing interindividual distance; (3) attempting to establish a mount series; and (4) maintaining a mount series. Signals recorded in the courtship context were then examined, categorical form by categorical form, with respect to the details of form and their distributions across stages and across troops.

Throughout study of the films and notes, I attempted to take account of the effects of memory on human perception [NORMAN, 1969] and, therefore, to anchor the analysis on the intersubjective records of signals on the film rather than on the descriptions of them in the notes. I believe that this caution fostered a more critical review and comparison of signal forms across troops than a study of notes alone, skewed by my changing perceptions in the field, would have allowed. During review of the films, however, the descriptions served to direct my attention to distinguishing features of signals that were not always as readily discernible in the films alone. Details of the categories, sorting, and study of the data have been described elsewhere [STEPHENSON, 1973]; and detailed comparisons of signal forms in selected contexts are in preparation, together with study films which will be available for review upon request.

Five Examples of Group-Specific Communication Patterns

To illustrate some of the findings and some of the problems to date, the forms of five patterns and their distributions across troops are described and discussed. The five patterns are labeled: male slide off, female lie back to, lip quiver, whirl pivot, and female mount male. The number of times the patterns were recorded in each stage of courtship and the number of encounters in which they were observed in each stage are presented in table II, along with the social distributions of observed performances of the patterns.

1. Male Slide off

The male slide-off pattern is a stylized dismount in which the male slides off to the female's side rather than lifting his torso and stepping back away from the female's raised hips to sit behind her. From his position alongside of

Table II. Observed frequency and social distribution of performances of five signal forms

Signal form	Troop	Total	A. Performances/encounter by stage				B. Performers by sex and discrete pairs			C. Discrete pairs by stage			
			1	2	3	4	♂	♀	prs	1	2	3	4
♂ slide off	M	103	0	0	22/4	81/12	5	5	10	0	0	3	7
	A	0	0	0	0	0	0	0	0	0	0	0	0
	K	3	0	2/2	1/1	0	1	3	3	0	2	1	0
♀ lie back to	M	13	0	3/2	6/3	4/3	4	4	7	0	2	3	3
	A	0	0	0	0	0	0	0	0	0	0	0	0
	K	0	0	0	0	0	0	0	0	0	0	0	0
Lip quiver	M	42	3/3	24/12	9/3	6/4	6	10	19	3	11	3	2
	A	0	0	0	0	0	0	0	0	0	0	0	0
	K	34	2/2	28/23	4/2	0	5	12	18	2	17	1	0
Whirl pivot	M	8	3/3	5/5	0	0	4	6	7	3	5	0	0
	A	71	1/1	70/30	0	0	9	17	25	1	24	0	0
	K	11	1/1	9/9	1/1	0	6	7	12	1	9	1	0
♀ mount ♂	M	80	0	5/3	75/9	0	4	4	7	0	2	6	0
	A	120	0	2/1	118/16	0	4	6	9	0	1	8	0
	K	0	0	0	0	0	0	0	0	0	0	0	0

M = Miyajima; A = Arashiyama; K = Koshima.

the female, the male then grabs the nape of the female's neck or the hair on her shoulder or back and holds her at arm's length while staring at her head. Usually the female stares back, and mutual eye contact is made. Often either or both vocalize repeatedly. The male may hold the female for a few seconds or up to a minute or more, then release her and resume sitting behind her.

Performances of male slide off were seen regularly in stages 3 and 4 at Miyajima, were not seen at Arashiyama, and were rarely seen at Koshima. The social distribution of this pattern at Miyajima suggests that it is group-specific. The proportions of males, females, and discrete pairs involved in performance of this pattern (table II B) are not significantly different from the proportions of males, females, and discrete pairs observed during the study period there (table I B; $\chi^2 = 4.464$; d.f. $= 2$; $p > 0.10$). That is, the observed social distribution of performances of male slide off is representative of the distribution of all males, females, and discrete pairs observed at Miyajima. Only one male, Kaminari, the old leader (more than 30 years old and thought to be senile) [M. KAWAI, personal communication, 1969], performed this pattern at Koshima and never in stage 4. Indeed, Kaminari was never observed in stage 4 during the study period, i. e., he did not attain a mount series with a female, although he frequently ejaculated from masturbation. Dr. KAWAI regarded Kaminari as 'impotent'. I regard his rare performances of male slide off as an idiosyncratic mannerism, the appearance of which is very similar to male slide off as performed by Miyajima males.

The problem inherent in this example is one of setting social criteria for characterizing a pattern as group-specific. One way in which a pattern can arise in a group is by members imitating the performance or the usage of a pattern previously peculiar to one of their fellows. Furthermore, a change in usage could affect the frequency of its performance by members of the group. For example, masturbation to ejaculation was observed frequently and was performed by all leader and subleader males in the study period at Koshima in 1968–1969. Only one instance was observed at Arashiyama, and none was observed at Miyajima. Prior to the 1965–1966 mating season at Koshima, however, masturbation to ejaculation was as rare as it was at the other study troops in 1968–1969. In 1965–1966, Kaminari began to masturbate to ejaculation regularly. In 1966–1967, the first (Kaminari) through third ranking males were regularly doing so, and in 1967–1968, the first through fifth ranking males were doing so [S. MITO, personal communication, 1969]. Over these four mating seasons, masturbation to ejaculation spread in social distribution through the higher classes of males in the troop. Although the leader and the subleader classes of males could be characterized as regularly masturbating to ejaculation in 1968–1969, they could not have been so characterized in 1966–1967.

To examine whether the differential social distribution of male slide off in the courtship context across troops is due to sampling differences across troops, the numbers of discrete pairs recorded in courtship encounters at high and middle quality levels of observation are presented by stage of courtship in table III. All courtship encounters were evaluated for quality independently of the search for incidences of particular signal forms. The quality of an encounter was classified as high when it was recorded on film or lasted longer than two minutes and was described in detail. Quality was classified as middle level when an encounter was less than two minutes long but was described in detail. If the occurrence of an encounter was simply noted with little or no detailed description, its quality was classified as low. Only encounters of high

Table III. Numbers of discrete pairs in courtship stages 1–4 observed at high and middle quality levels

	High				Middle			
	1	2	3	4	1	2	3	4
Miyajima	11	25	14	12	14	22	3	6
Arashiyama	15	44	16	14	24	35	8	5
Koshima	6	33	19	10	9	35	12	4

and middle quality where the identity of male and female is known have been considered here. The proportions of discrete pairs observed at middle and high quality levels are not significantly different across troops ($\chi^2 = 16.664$; d.f. = 14; p > 0.20); hence the observed differences in social distribution of male slide off (and of the other patterns as well) are probably not due to differences in sampling intensity.

2. Female Lie Back to

On occasion at Miyajima, while the male held the female at arm's length after dismounting by sliding off, the male would roughly push the female away, causing her to fall downslope or off of the rock or log on which they were courting or consorting. In returning to the male in this apparently emotionally tense situation, the female would approach very hesitantly, and just before arriving, would turn and lie on her side with her back to him.

Female lie back to is an example of the case of signals similar, contexts different (p. 60 above). Lying with back to was frequently performed by females (and males) in groom solicitation in all study troops; but although females were also occasionally pushed away by males in courtship at Arashiyama and Koshima, only at Miyajima was the female lie back to pattern used in the courtship context. Expected values are too low (less than 2.0) to compare the distribution of males, females, and pairs involved in the performance of female lie back to in the courtship context at Miyajima with the proportions of males, females, and pairs observed there during the study.

3. Lip Quiver

During performance of the lip quiver pattern, puckered lips remain together and move up and down over the teeth of the sender at about 12 cycles per second. The usual posture of the sender is a quadrupedal stance held with head and shoulders lowered and the head rotated back on the axis with jaw thrust forward such that the face appears prognathic. In courtship, the sender was typically a male and the receiver was typically a female. The sender was usually oriented toward the receiver [as addressee; SEBEOK, 1965], but at Miyajima, the pattern was sometimes performed while the male's body was oriented away from the female with the male looking back over his shoulder at her as he passed by or moved away from her.

Performances of lip quiver were recorded in all four stages of courtship at Miyajima, were not recorded in any stages of courtship at Arashiyama, and were recorded in stages 1–3 at Koshima. The proportions of males, females, and discrete pairs involved in the performance of lip quiver in the courtship context at Miyajima (table II B) are not significantly different from the proportions of males, females, and discrete pairs observed during the study period there (table I B; $\chi^2 = 2.455$; d.f. = 2; p > 0.20). Similarly at Koshima, the proportions are not significantly different ($\chi^2 = 4.467$; d.f. = 2; p > 0.10).

The dual social distribution of lip quiver at Miyajima and Koshima presents another problem for the notion of group-specific. Although Arashiyama males were not observed to perform lip quiver in the courtship context, they were able to perform the pattern *per se*. Lip quiver was sometimes performed in encounters between a male and the progeny of a female in the course of the male's efforts to drive the progeny out of the vicinity of the female while he courted or consorted with her. Miyajima and Koshima males also used lip quiver in this manner and context. This is similar to the case of signals similar contexts different for the pattern, female lie back to. As with lip quiver, members of more than one group are known to be capable of performing that pattern; but female lie back to is used in the courtship context only by Miyajima members. Lip quiver is used in the courtship context by Miyajima and by Koshima members, but it is not used in courtship by Arashiyama members. Until further data indicate otherwise, I regard lip quiver as a group-specific pattern specific to two distinct groups.

4. Whirl Pivot

Whirl pivot is a highly stylized approach and withdrawal of a male toward and away from a female. The male approaches with head and shoulders low and head rotated back on the axis with jaw thrust forward such that the face appears prognathic, a posture much like that associated with lip quiver. Between 15 and 5 ft from the female, the male momentarily freezes in the posture, then quickly wheels around 180 degrees by pivoting on his hind feet, and walks away.

Performances of this basic movement were recorded in stages 1 and 2 at all three study troops and once in stage 3 at Koshima. When one compares the social distribution of observed performances (table IIB) with subjects observed (table IB), significantly more males and more females were involved in the performance of whirl pivot at Arashiyama ($\chi^2 = 7.699$; d.f. $= 2$; p < 0.05) and Koshima ($\chi^2 = 6.673$; d.f. $= 2$; p < 0.05) than expected. Expected values were too low (less than 2.0) to make this comparison in the Miyajima data. The difference across troops in the use of this basic movement is that at Miyajima and Koshima it was performed only once in the course of an encounter, but at Arashiyama it was performed in series of two or more times in 14 of 31 encounters. The longest series of whirl pivots recorded at Arashiyama was seven times in seven minutes, after which the male attempted to establish a mount series with the female. Differences in the frequencies of singletons and series of whirl pivots across troops are significant ($\chi^2 = 11.918$; d.f. $= 2$; p < 0.01).

The different distributions of whirl pivots in singletons and series across troops points out one of the problems in the delineation of a pattern by an observer. With reference to the notion of a pattern as a recurrent perception, i.e., as a perceived type or kind, I regard the serial aspect of performance of this pattern by Arashiyama males as a group-specific pattern since I came to expect performances in series during the study period at Arashiyama, but expected only single incidences of whirl pivot during the study periods at Miyajima and Koshima. In this sense, although the basic movement setting up the signal has dimensions in both space and time, pattern *per se* is a dimensionless notion.

5. Female Mount Male

In female mount male, the female may mount the male's hips in male mount female fashion, with her hind feet clasped behind the knees or on the back of the thighs of the male's corresponding hind legs; or the female may sit in the saddle of the male's back. If the male has not rocked forward from sitting into the quadrupedal present posture, the female may mount his shoulders instead of his hips or back. A female mounted on a presenting male may thrust with her pelvis or may rub her crotch forward and backward on the male's back, or if he is sitting, may rub her crotch up and down on his neck.

Female mount male was performed, usually in series, in stages 2 and 3 by females at Miyajima and Arashiyama; but its performance was not observed at Koshima. Indeed, over the years of surveillance by Japanese primatologists since studies were initiated there in 1949, female mount male has never been observed at Koshima [M. KAWAI, personal communication, 1969; J. ITANI, personal communication, 1972]. Expected values are too low (less than 2.0) to compare the distributions of males, females, and pairs performing to males, females, and pairs observed at Miyajima and at Arashiyama.

Detailed consideration of the within-troop social distribution of performances of female mount male brings out another problem in the elucidation of group-specific communication patterns. At Miyajima, only females in the lower half of the social hierarchy were observed mounting males; and except for one case in which the third ranking male was mounted twice by a female, only subleader males were observed being mounted by females. At Arashiyama, females observed mounting males ranged in rank from first to thirty-fifth; but only the first through fourth ranking males were observed being mounted by females. In light of the capacity for complex-dependent rank relations in the Japanese macaque, at the present refined level of contextual control, female mount male appears to be used differently at Miyajima than at Arashiyama. Hence, each usage can be taken as a group-specific communication pattern, since different signs relate sign vehicles residing in the similar signals set up in the course of female mount male to different contexts in the two troops. This is similar to the case among human beings in which a particular mode of address (e. g., 'Sir') can be used by members of one group when addressing peers and by members of another only when addressing superiors.

Rules of Usage

In some of the recent literature concerning animal communication [e. g., MARLER, 1961, p.229; SMITH, 1965, p.406], the pragmatic dimension of semiotic has been used somewhat interchangeably with the semantic dimension; and the notion of context has thereby become confused with the notion of 'meaning'. In SMITH's [1968, p.48] analysis of animal communication, the sender's semiosis is considered in terms of 'messages' and the receiver's semiosis in terms of 'meaning'. Messages refer to things being sent and meanings refer to things being received. In terms of the previous discussion here, SMITH's usage of message and meaning emphasizes the fact that although a sender sets up a sign vehicle *per se*, the sign vehicle is received as part of the immediate context of the receiver. The effect of being imbedded in and necessarily sorted from immediate context may yield an interpretant as it is produced in the receiver that is different from the interpretant as it is represented in the sender. Although this distinction is logical and important to recognize, the outcomes of communication events are predictable by an observer; hence it can be assumed that the sender and receiver have common alphabets of signs and that the receiver is able appropriately to sort sign vehicles from immediate context. On the other hand, when one treats as referent the set of contextual items that correlates with usage of a pattern setting up the signal in which the sign vehicle resides and treats as response the outcome of the communication event, the observer's insight into the relation between the interpretant as it is represented in the sender and the interpretant as it is produced in the receiver reduces to an evaluation of the congruity of referent and response. The observer essentially sorts to make sense of the relation between apparent referent and apparent response in time, surely a matter of taste on the part of the observer. Sorting signals from context and sign vehicles from signals is akin to the figure-ground problem. By controlling context, the configurations of sign vehicles can be delineated from the background of immediate context. At a more general level, 'meaning' is a term that MORRIS [1938, p. 48] specifically noted '... adds nothing to the set of semiotical terms'. MORRIS clarified this opinion by stating [p. 47]: 'Since the meaning of a sign is exhaustively specified by the ascertainment of its rules of usage' (or in the above terms, by the rules of usage of its sign vehicle), 'the meaning of any sign is in principle exhaustively determinable by objective investigation'.

Ascertaining the rules of usage of sign vehicles, rather than ascribing 'meaning' to their signs, facilitates the analysis of nonhuman communication events in two ways. At the observer level, the notion of a rule of usage takes better account of the observer's position outside of the channel he is observing. For the observer, the rule of usage underlying the incidence of a particular sign vehicle is experienced as a particular correspondence between a particular form of signal in which the sign vehicle resides and the context of observation in which a particular referent is apparent. This heuristic position emphasizes the legisign aspect of PEIRCE's notion of sign [cf. HARTSHORNE and WEISS, 1931–1935, vol. 2, § 246] in the observer's semiosis and treats the correspondence *per se* as focal point for the semiotical analysis of the communication event he is observing. At the user level, an emergent feature of the pragmatic dimension of semiotic can be more fully taken into account: the fact that the sign aspect of the semiotical triad is not necessarily coupled to the signal *per se* in a communication event, but that through usage, aspects of signals can come to function as sign vehicles in their carrying of relations between signs and designata that mediate relations between signs and interpretants. Rules of usage take better account of the social acquisition of arbitrary signs.

Summary

Japanese macaques have the capacity for social acquisition of arbitrary signs. This capacity could be manifest in their communication system as a duality in signal-context relations. Semiotic, the theory of signs, provides a conceptual framework for taking account of the duality in field studies of nonhuman communication systems. Some of the problems in testing for group-specific communication patterns in a field study of three troops of Japanese macaques are discussed. Five examples are presented and discussed in terms of rules of usage. Given the observer's position outside of the communication system he is observing, ascertaining rules of usage seems more heuristic than searching for the meanings of signs in studies of nonhuman communication systems.

References

ALTMANN, S. A.: Social behavior of anthropoid primates: analyses of recent concepts; in BLISS Roots of behavior, pp. 277–285 (Harper, New York 1962).

ALTMANN, S. A.: Sociobiology of rhesus monkeys. II. Stochastics of social communication. J. theoret. Biol. *8:* 490–522 (1965).

CHANCE, M. R. A. and MEAD, A. P.: Social behavior and primate evolution. Symp. Soc. exp. Biol. *7:* 395–439 (1953).

CHERRY, C.: On human communication (The MIT Press, Cambridge 1966).

CROOK, J. H.: Social organization and the environment: aspects of contemporary social ethology. Anim. Behav. *18:* 197–209 (1970a).

CROOK, J. H.: The socio-ecology of primates; in CROOK Social behavior in birds and mammals (Academic Press, London 1970b).

DELGADO, J. M. R.: Evolution of physical control of the brain. James Arthur Lecture on the Evolution of the Human Brain (The American Museum of Natural History, New York 1965).

ETKIN, W.: Cooperation and competition in social behavior; in ETKIN Social behavior and organization among vertebrates, pp. 1–34 (University of Chicago Press, Chicago 1964).

FRISCH, J. E.: Individual behavior and intertroop variability in Japanese macaques; in JAY Primates: studies in adaptation and variability, pp. 243–252 (Holt, Rinehart & Winston, New York 1968).

GARTLAN, J. S. and BRAIN, C. K.: Ecology and social variability in *Cercopithecus aethiops* and *C. mitis;* in JAY Primates: studies in adaptation and variability, pp. 253–292 (Holt, Rinehart & Winston, New York 1968).

GOUDGE, T. A.: The thought of C. S. Peirce (Dover Publications, New York 1969).

HALL, K. R. L.: Observational learning in monkeys and apes. Brit. J. Psychol. *54:* 201–226 (1963a).

HALL, K. R. L.: Variations in the ecology of the chacma baboon, *Papio ursinus*. Symp. zool. Soc., Lond. *10:* 1–28 (1963b).

HALL, K. R. L.: Social learning in monkeys; in JAY Primates: studies in adaptation and variability, pp. 383–397 (Holt, Rinehart & Winston, New York 1968).

HALL, K. R. L. and GOSWELL, M. J.: Aspects of social learning in captive patas monkeys. Primates *5:* 59–70 (1964).

HANSEN, E. W.: The development of maternal and infant behavior in the rhesus monkey; Ph. D. Diss., Univ. of Wisconsin (1962).

HARLOW, H. F.: The formation of learning sets. Psychol. Rev. 56: 51–65 (1949).

HARTSHORNE, C. and WEISS, P.: Collected papers of Charles Sanders Peirce (Harvard University Press, Cambridge 1931–1935).

HOCKETT, C. F.: A course in modern linguistics (MacMillan, New York 1958).

HOCKETT, C. F. and ALTMANN, S. A.: A note on design features; in SEBEOK Animal Communication, pp. 61–72 (Indiana Universita Press, Bloomington 1968).

HUBEL, D.: The visual cortex of the brain. Scientific American 209 (5): 54–62 (1963).

HUBEL, D. and WIESEL, T. N.: Receptive fields of cells in striate cortex of very young, visually inexperienced kittens. J. Neurophysiol. 26: 994–1002 (1963).

HUXLEY, J.: A discussion on ritualization of behaviour in animals and man. Phil. trans. roy. soc. London B. 251: 247–408 (1966).

IMANISHI, K.: Social organization of subhuman primates in their natural habitat. Curr. Anthropol. 1: 393–407 (1960).

JAY, P.: Primates: studies in adaptation and variability (Holt, Rinehart and Winston, New York 1968).

KAWAI, M.: Newly-acquired pre-cultural behavior of the natural troop of Japanese monkeys on koshima island. Primates 6 (1): 1–30 (1965).

KAWAI, M.: A case of unseasonable birth in Japanese monkeys. Primates 7: 391–392 (1965).

KAWAMURA, S.: The matriarchal social order in the minoo-B-troop – a study on the rank system of Japanese macaques. Primates 1 (2): 149–156 (1958).

KAWAMURA, S.: The process of sub-cultural propagation among Japanese macaques. Primates 2 (1): 43–54 (1959).

KOFORD, C. B.: Group relations in an island colony of rhesus monkeys; in SOUTHWICK Primate social behavior, pp. 136–152 (D. Van Nostrand Co., Princeton 1963).

KONORSKI, J.: Integrative activity of the brain (University of Chicago Press, Chicago 1967).

KUMMER, H.: Two variations in the social organization of baboons; in JAY Primates: studies in adaptation and variability (Holt, Rinehart & Winston, New York 1968).

LENNEBERG, E. H.: Biological foundations of language (John Wiley & Sons, New York 1967).

LETTVIN, J. Y.; MATURANA, H. R.; MCCULLOUGH, W. S., and PITTS, W. H.: What the frog's eye tells the frog's brain. Proc. Inst. Radio-Engineers. 47: 1940–1951 (1959).

MARLER, P.: The logical analysis of animal communication. J. theoret. Biol. 1: 295–317 (1961).

MARLER, P.: Communication in monkeys and apes; in DEVORE Primate behavior: field studies of monkeys and apes, pp. 544–584 (Holt, Rinehart & Winston, New York 1965).

MASON, W. A. and HOLLIS, J. H.: Communication between young rhesus monkeys. Anim. Behav. 10: 211–221 (1962).

MILLER, R. E.: Experimental approaches to the physiological and behavioral concomitants of affective communication in rhesus monkeys; in ALTMANN Social communication among primates, pp.125–134 (University of Chicago Press, Chicago 1967).

MILLER, R. E.; BANKS, J. H., jr., and KUWAHARA, H.: The communication of affects in monkeys: cooperative reward conditioning. J. genet. Psychol. 108: 121–134 (1966).

MILLER, R. E.; MURPHY, J. V., and MIRSKY, I. A.: Relevance of facial expression and posture as cues in communication of affect between monkeys. Arch. gen. Psychiat. 1: 480–488 (1959).

MITCHELL, G.: Persistent behavior pathology in rhesus monkeys following early social isolation. Folia primat. *8:* 132–147 (1968).

MORRIS, C.W.: Foundations of the theory of signs. Int. Encycloped. unif. Sci. *1 (2):* 1–59 (1938).

MORRIS, C.W.: Signs, language and behavior (Braziller, New York 1946).

MORRIS, C.W.: Signification and significance: a study of the relations of signs and values (The MIT Press, Cambridge 1964).

NAPIER, J.R. and NAPIER, P.H.: A handbook of living primates (Academic Press, New York 1967).

NORMAN, D.A.: Memory and attention: An introduction to human information processing (John Wiley & Sons, New York 1969).

OPLER, M.E.: The human being in culture theory. Amer. Anthropol. *66:* 507–528 (1964).

SACKETT, G.P.: Monkeys reared in isolation with pictures as visual input: evidence for an innate releasing mechanism. Science *154:* 1468–1473 (1966).

SEBEOK, T.A.: Animal communication. Science *147:* 1006–1014 (1965).

SENKO, M.G.: The effects of early, intermediate, and late experience upon adult macaque sexual behaviour; M.S. thesis, Univ. of Wisconsin (1966).

SKINNER, B.F.: The phylogeny and ontogeny of behavior. Science *153:* 1205–1213 (1966).

SMITH, W.J.: Message, meaning, and context in ethology. Amer. Natural. *99:* 405–409 (1965).

SMITH, W.J.: Message-meaning analysis; in SEBEOK Animal communication: techniques of study and results of research, pp. 44–60 (Indiana University Press, Bloomington 1968).

STEPHENSON, G.R.: Cultural acquisition of a specific learned response among rhesus monkeys; in STARCK, SCHNEIDER and KUHN Progress in primatology, pp. 279–288 (Fischer, Stuttgart 1967).

STEPHENSON, G.R.: The influence of a group bonding on geographical localization. Biologist *50:* 126–141 (1968).

STEPHENSON, G.R.: Testing for group-specific communication patterns in Japanese macaques; Ph.D. Diss., Univ. of Wisconsin (1973).

STEPHENSON, G.R.: Communication and population structure; in CARPENTER Behavioral regulators of behavior in primates (revision of presentation as The biology of communication and structure: a model of primate populations). 8th Int. Congr. Anthrop. Ethnol. Sci., Kyoto 1968 (Burchnell University Press, in press).

SUZUKI, A.: On the insect-eating habits among wild chimpanzees living in the savanna woodland of eastern Tanzania. Primates *7:* 481–487 (1966).

THOMPSON, N.S.: Some variables affecting the behavior of irus macaques in dyadic encounters. Anim. Behav. *15:* 307–311 (1967).

TINBERGEN, N.: The study of instinct (Clarendon Press, Oxford 1951).

TINBERGEN, N.: 'Derived' activities; their causation, biological significance, origin, and emancipation during evolution. Quart. Rev. Biol. *27:* 1–32 (1952).

WYKOFF, W.: Semiosis and infinite regressus. Semiotica *2:* 59–67 (1970).

ZUSNE, L.: Visual perception of form (Academic Press, New York 1970).

Author's address: Dr. GORDON R. STEPHENSON, Department of Zoology, University of Wisconsin, *Madison, WI 53706* (USA)

Symp. IVth Int. Congr. Primat., vol. 1: Precultural Primate Behavior,
pp. 76–87 (Karger, Basel 1973)

Age Changes of the Vocalization in Free-ranging Japanese Monkeys

A. NISHIMURA

Primate Research Institute, Kyoto University, Inuyama, Aichi

Introduction

Since 1965, I have been collecting quantitative data in order to analyze
the ontogeny of behavior in free-ranging Japanese monkeys. There has been
little research of this type on Japanese monkeys in spite of the large body of
sociological data gathered over the past 20 years. It is important to study
quantitatively age changes in behavior in order to understand the socialization
process of individual monkeys and in order to facilitate comparisons with the
research of various workers. Techniques for observing and scoring behavior
have been developed by American and European investigators who have
studied the behavior of primates in captivity. I have tried to introduce these
techniques into my work as much as possible.

This paper will describe age changes in the vocalizations of free-ranging
Japanese monkeys during their first four years. This is only a part of my work;
the development of behavior in general will be reported elsewhere. Vocalization of Japanese monkeys has been studied intensively by ITANI [1963],
KAWABE [1965], and TAKEDA [1965, 1966]. ITANI has described and classified
the repertoire and has discussed the function that vocalization plays in interindividual relationships and in group integration in the natural life of Japanese
monkeys. The other two investigators have studied the development of
vocalization in hand-reared infants for their first 30 weeks. Thus, there are
marked differences between the material studied by ITANI and by KAWABE and
TAKEDA, i.e. free-ranging adults, on the one hand and hand-reared infants,
on the other. My work will partly close the gap between these two studies. At
the same time, it is expected that the quantitative approach used here, but not

by these others, will provide more understanding of the ontogeny of vocal behavior.

Material and Method

All data used for analysis were obtained around the feeding ground of Takasakiyama. Of the two methods used for the study of development, the cross-sectional and the longitudinal, I used the former, although the latter is more frequently used in the study of behavior development in non human primates [e.g. HINDE et al., 1964; HANSEN, 1966; JENSEN et al., 1967]. At Takasakiyama, the cross-sectional method is more feasible. For example, more than 200 infants are born every year [MASUI et al., in press] and the birth period covers five months, from May to September [KAWAI et al., 1967].

Material used for analysis in this paper consists mainly of 79 sample observations made on 72 individual monkeys (seven of the monkeys were observed at two different times). Each sample observation is the record of behavior on a check-sheet divided into half-minute periods and containing the continuous record of an individual for the total time it was observed. About half of the sheet was left blank for the writing of qualitative notes and for recording items that were not completely categorized or did not occur frequently. Since vocalizations had not been well categorized at the time of collecting the data, much of it was recorded by a combination of Japanese letters or alphabet in the blank part of the sheet. Table I shows the age categories and the number and size (i.e., number of minutes of observation) of the samples. The age categories are given in months (30 days) after birth, with a range of −5 to +10 days around each value. The size of the observation samples taken during and after 1969 is about 100 minutes per sample, but some samples taken before 1969 greatly exceed this.

Vocal Repertoire and Situation

The vocalizations contained in the 79 sample observations were classified into 11 elementary types, principally according to sound quality. Of these, nine types seem to belong to the classification 'harsh noises', and two to 'clear calls' [ROWELL and HINDE, 1962]. Each type of vocalization is usually emitted by itself, but sometimes two or three are emitted in combination. Cor-

Table I. The number and size (in minutes of observation) of samples in each age category

	Months																Total
	0.5	1	2	3	4	5	6	7	8	9	10	11–15	18–21	24–29	36–39	48–51	
Number																	
Males	0	3	2	2	2	2	2	1	2	2	2	4	2	3	3	3	35
Females	2	3	2	3	2	2	2	2	4	1	2	4	3	5	3	4	44
Total	2	6	4	5	4	4	4	3	6	3	4	8	5	8	6	7	79
Size, min																	
Min.	100	100	98	100	102	102	106	94	97	101	103	101	93	90	90	95	
Max.	100	103	212	170	128	112	111	111	114	106	109	211	103	166	362	250	
Mean	100	102	129	116	109	110	110	102	104	103	105	149	101	120	169	136	

respondence of each type with the vocalizations of Japanese monkeys in other studies, or with the vocalizations of other macaques, is indicated in brackets at the end of the description of each type. References frequently cited will be referred to by number: (1) TAKEDA [1965, 1966], (2) ITANI [1963], (3) ROWELL [1962], and (4) ROWELL and HINDE [1962]. In the descriptions that follow, the method for expressing the sound by letter combination is that of ITANI [1963].

A. Harsh Noises (Types 1–9)

1. Short Cry (⟨k⟩ ⟨ki (t)⟩ ⟨gi (t)⟩ ⟨kyat⟩ ⟨gyat⟩ ⟨kyot⟩ ⟨gyot⟩)
Each sound is short. The cries are emitted alone or in series. This sound occurs in such mother-infant interactions as searching for the nipples of the mother, abrupt violent body movement of the mother (whom the infant then clutches), the mother's moving the infant off from her, and contact deterrence or breaks by mother [KAUFMAN and ROSENBLUM, 1966]. The short cry is also emitted in agonistic interactions with individuals other than the mother. However, the direct cause of this sound was uncertain in many cases [Stock A and B (1); F-5, F-6, and B-4 (2); gecker (4) and geckering screech (3)].

2. Long Cry ⟨kiyaa⟩ ⟨giyaa⟩ ⟨kii⟩
This sound is longer than the short cry and emitted more frequently in series than as a single sound. Most of these long cries are made during inter-individual antagonism, where the emitter usually takes a defensive posture [Stock C (1); B-1 (2); screech and scream (3, 4)].

3. Aggressive Cry ⟨ga (t)⟩ ⟨go (t)⟩ ⟨gaa⟩ ⟨garr⟩
This strong, usually short sound occurs alone or in series. Aggressive cries are without exception emitted during interindividual antagonism, where the emitter almost always takes an aggressive posture [C_3, C_{3-1} (1); C-1, C-2, C-3 (2); bark, roar, growl and pant threat (3, 4)].

4. Tantrum ⟨aga.ga.ga...⟩ ⟨agya.gya.gya...⟩
Each syllable resembles the short cry but is shorter and stronger and is always emitted in series. The tantrum is peculiar to infants. In this study, it occurred in individuals from two to ten months old. This sound is always accompanied by head and body shaking. When infants are emitting it, it appears that they are anticipating maternal care [C_1 and C_{1-1} (1); weaning tantrum (?): HINDE, et al., 1964].

5. ⟨oji.ji.ji...⟩ ⟨ktyu.tyu.tyu...⟩

This noise is difficult to write as a 'word'. It is emitted with a small mouth opening, and the ending of a 'word' is obscure and sometimes indistinguishable from a clear call. Situations were almost the same as in the simple calls (10, below) [F-4 (?) (2)].

6. ⟨krr⟩ ⟨grr⟩

This sound resembles a simple call (10, below), except for the insertion of a rolled 'r' sound. It was used as a greeting by a 7-month-old male and a 19-month-old male, but in other cases the context was uncertain [greeting growl (?): HINDE, *et al.*, 1964].

7. ⟨gu⟩ ⟨guu⟩

Emitted as a single sound, this noise sometimes follows the simple call. It was used as a greeting in several cases. But again, the context was often unclear [A-3, A-4 (2); food bark (?) (4)].

8. Shrill Bark ⟨kuon⟩

This sound resembles the short cry, but is shorter, stronger and higher pitched. Only one case was observed in this study, where it was emitted by a 24-month-old male who became aware of a dog near the troop [group D (2); shrill bark (3, 4)].

9. Girning ⟨ngaa⟩

A low, and calm noise, accompanied by nasal sounds, only one case of girning was observed in this study. The sound was made by a 36-month-old female; the context was uncertain. At Takasakiyama, I observed some adults emitting this noise rather frequently at workers, apparently when asking for food [A-13 (2); girning (4)].

B. Clear Calls (Types 10 and 11)

The number of clear calls, collected in my samples, based on my notation system, was about 40. However, clear calls are quite variable. They vary among individuals in what appears to be the same situation [ROWELL and HINDE, 1962; TAKEDA, 1965, 1966]. Also clear calls readily grade into one another. I have tentatively classified clear calls as being of two types: simple and complex. It may be possible to classify the complex calls further, depend-

1, 2

Fig. 1. One-year-old female emitting simple call.
Fig. 2. One-year-old male emitting complex call.

ing on a detailed contextual analysis. Such investigation is now being conducted.

10. Simple Call ⟨hu⟩ ⟨hū⟩ ⟨ku⟩ ⟨nku⟩ ⟨ho⟩ ⟨hō⟩

A short and low-pitched call, sometimes having a harsh sound in the infant, this sound is emitted with a small mouth-opening (fig. 1). The simple call is metaphorically referred to as 'unintentionally uttered murmurs' [ITANI, 1963] or 'sobbing' [TAKEDA, 1965, 1966]. The infant usually makes this call without any accompanying behavior, such as aggressive, defensive, play, or other social behavior. Usually, it is emitted when the infant is by itself [A-1 (2), stock D (?) (1)].

11. Complex Call ⟨hui⟩ ⟨huii⟩ ⟨huui⟩ ⟨kyuii⟩ ⟨pyuii⟩ ⟨huo⟩ ⟨hyaa⟩ ⟨kyuaa⟩ ⟨hiyuu⟩

This is a long and high-pitched call (fig. 2). The context was clear in some cases (e. g., calling mother, response to calls emitted by others); however, in many cases, it was not [A-2, -7, -8, -9, -10, -11, -12 (2); stock E (1)].

C. Compound Sounds

Each type of vocalization listed above is usually emitted by itself. But sometimes two different types of vocalization occur in series. In particular, two types of clear calls sometimes occur sequentially (see fig. 5). Clear calls also occurred with the following harsh noises: short cry and type '5', '6', or '7'. Harsh noises did not sequentially combine with each other, although the long cry sometimes was abruptly followed by the aggressive cry in the same situation, or vice versa [ITANI, 1963].

Age Changes in the Frequency of Vocalization

The frequency of vocalizations was calculated as the percent of the total of half-minute intervals of observation in which they were recorded (fig. 3–5).

1. Total Vocalization
The frequency of total vocalization increases rapidly with age for the first half-year of life and then begins to decrease, falling to a low level by the start of the second year. It does not change much, or only slowly decreases, after that time (fig. 3).

2. Short Cry (fig. 4)
This sound is emitted in 2% of the half-minute intervals observed in the middle of first month. The frequency increases to 4.5% at the beginning of three months, then decreases to 2% at one and one-half years, until reaching a stable level at 0.3% by the beginning of the second year. It is thus quite different at one and one-half years and at the beginning of the second year.

3. Long Cry (fig. 4)
This sound was not heard at all in the sample of animals less than a half-month of age. However, KAWABE [1965] and TAKEDA [1965, 1966] report that this type of sound is first emitted within the first two weeks of life. In my observations, it appeared in the samples of the one-month age category in less than one per 100 minutes on an average. It seems to increase after that until one or one and one-half years, where it appears in 2% of the half-minute periods, and then decreases slowly to 0.5% at the beginning of the fourth year. The long cry is, therefore, emitted less frequently than the short cry from birth to one and one-half years and more frequently after that.

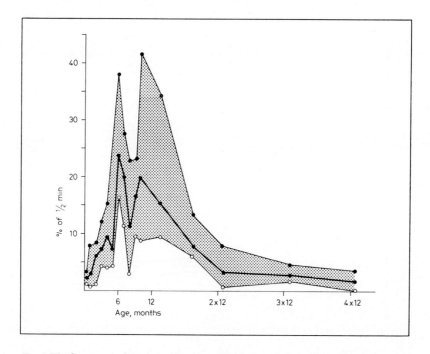

Fig. 3. The frequency of total vocalization calculated as percent of half-minute intervals in which any kind of vocalization was emitted. Thick line shows average; shaded area, the range.

4. Aggressive Cry (fig. 4)

This was not heard in samples taken during the first nine months, except during the sixth month, where it appeared only once. After nine months, it appeared in all the samples, although the amount is small, usually in less than 0.5% of the half-minutes. KAWABE [1965] and TAKEDA [1965, 1966] have reported that the first appearance of this type of vocalization (in the 24th week for the former and the 11th week for the latter) is later than the short cry or the long cry. The aggressive cry is emitted less frequently than the short cry or long cry, until one and a half years. After two years, it is emitted more frequently than the short cry but still less frequently than is the long cry.

5. Other Harsh Noises

Harsh noises other than the three discussed separately above occurred much less frequently; the number of half-minutes in which they were emitted is, therefore, shown in table II in terms of raw data, rather than of relative

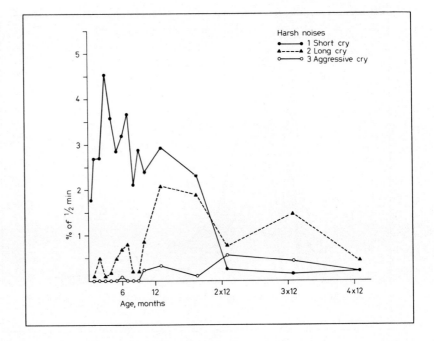

Fig. 4. The mean percent of half-minute intervals in which short cry, long cry, and aggressive cry were emitted

Table II. The number of half-minutes in which sound types '4', '5', '6', '7', '8' and '9' occurred in each age category. The figures in parentheses are the numbers of half-minutes in which they were emitted in combination with clear calls

	Months														
	0.5	1	2	3	4	5	6	7	8	9	10	11–15	18–21	24–29	36–39	48–51
Harsh noises																
4 (Tantrum)		1	1	1		1	2	1		2						
5 ⟨oji.ji...⟩		1		1	2		2	(1)	1	1	1					
6 ⟨krr⟩				(1)	1		2					1+(3)				
7 ⟨gu⟩						1			1+(1)				2+(3)	1		
8 (Shrill bark)													1			
9 (Girn)															1	

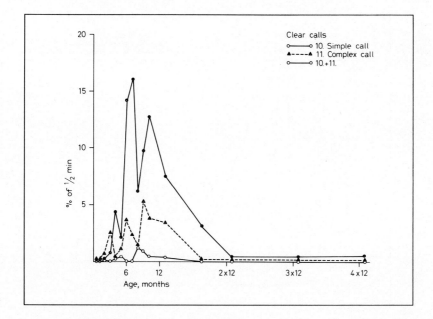

Fig. 5. The mean percent of half-minute intervals in which simple clear calls, complex clear calls, and compounds of these sounds were emitted.

frequency per age category. From the table it may be seen that the tantrum cry and the type '5' sound are emitted from two months to one year of age and that types '6', '7', '8' and '9' had already appeared in the seventh month, in the eighth month, at two years, and at three years, respectively.

6. Simple Call and Complex Call (fig. 5)

The change of frequency in the simple call with increasing age is quite dramatic. The proportion of half-minute periods in which the simple call is emitted is less than 0.5% for the first two months of life, but then it increases very rapidly to 16% at six or seven months. After a half-year it begins to decrease, so that by two years of age the frequency level is down to 0.5%, where it remains for the next two years.

Patterns of age-related change in the frequency of complex calls are almost the same as that for simple calls, except that the former are generally less frequent and attain a peak later. The frequency of compound sounds composed of simple and complex calls changes with age in the same pattern as the simple or complex call. However, the frequency of these particular

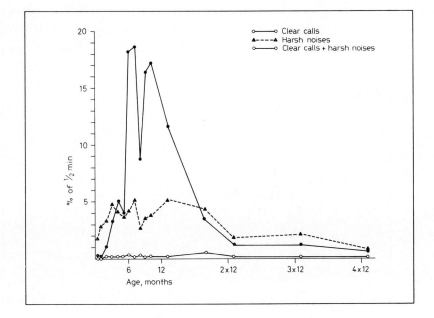

Fig. 6. The mean percent of half-minute intervals in which harsh noises, clear calls, and compounds of these sounds were emitted.

compound sounds is considerably smaller than either the simple or complex call.

7. Total Harsh Noises, Total Clear Calls, and Total Compound Sounds (fig. 6)
Relative frequencies were calculated for harsh noises, clear (simple and complex) calls, and compound sounds of both kinds (harsh and clear). Figure 6 shows that: (1) for the first two years, the frequency of clear calls changes more rapidly, and has a higher peak, than that of harsh noises; (2) harsh noises exceed clear calls in quantity only for the first three months and after one and a half years; and (3) for ages six months to one and one-half years, clear calls are emitted more frequently than harsh noises.

Summary and Conclusion

The behavior of free-ranging Japanese monkeys was observed over a seven-year period and was recorded on a standardized check sheet. In this paper, only the vocal behavior data for the first four years in the life of Japanese monkeys are analyzed. Vocalizations are first classified into 11 types, and then age-related changes in the frequency of vocal type presented. The results show that the vocalizations of Japanese monkeys change not only qualitatively but quantitatively in their first two years. We believe that more attention should be directed to vocal behavior and its ontogeny in infant and juvenile nonhuman primates. Such research might provide the comparative data needed for the study of the origins of human language.

References

HANSEN, E. W.: The development of maternal and infant behavior in the rhesus monkey. Behaviour 26: 107–149 (1966).

HINDE, R. A.; ROWELL, T. E., and SPENCER-BOOTH, Y.: Behavior of socially living rhesus monkeys in their first six months. Proc. zool. Soc., Lond. 143: 609–649 (1964).

ITANI, J.: Vocal communication of the wild Japanese monkey. Primates 4: 11–66 (1963).

JENSEN, G. D.; BOBBITT, R. A., and GORDON, B. N.: The development of mutual independence in mother-infant pigtailed monkeys, Macaca nemestrina; in ALTMANN Social communication among primates (University of Chicago Press, Chicago 1967).

KAUFMAN, I. C. and ROSENBLUM, L. A.: A behavioral taxonomy for Macaca nemestrina and Macaca radiata: based on longitudinal observation of family groups in the laboratory. Primates 7: 205–258 (1966).

KAWABE, S.: Development of vocalization based on the experiment of handreared Japanese monkeys. Kagaku Yomiuri 17 (6): 25–29 (1965).

KAWAI, M.; AZUMA, S., and YOSHIBA, K.: Ecological studies of reproduction in Japanese monkeys (Macaca fuscata). I. Problems of the birth season. Primates 8: 35–74 (1967).

MASUI, K.; NISHIMURA, A.; OSAWA, H., and SUGIYAMA, Y.: Population study of Japanese monkeys at Takasakiyama (1970–1971) (in press).

ROWELL, T. E.: Agonistic noises of the rhesus monkey (Macaca mulatta). Symp. zool. Soc., Lond. 8: 91–96 (1962).

ROWELL, T. E. and HINDE, R. A.: Vocal communication by the rhesus monkey (Macaca mulatta). Proc. zool. Soc., Lond. 138: 279–295 (1962).

TAKEDA, R.: Development of vocal communication in man-raised Japanese monkeys. I. From birth until the sixth week. Primates 6: 337–380 (1965).

TAKEDA, R.: Development of vocal communication in man-raised Japanese monkeys. II. From the seventh to the thirtieth week. Primates 7: 73–116 (1966).

Author's address: Dr. AKISATO NISHIMURA, Primate Research Institute, Kyoto University, Inuyama, Aichi 484 (Japan)

Symp. IVth Int. Congr. Primat., vol. 1: Precultural Primate Behavior,
pp. 88–101 (Karger, Basel 1973)

Influences of Phylogeny and Ecology on Variations in the Group Organization of Primates

J. S. GARTLAN

Institut de Recherches Médicales, Kumba, République Fédérale du Cameroun, and
Department of Psychology, University of Bristol, Bristol

Introduction

The relationship between social structure and habitat in primates depends partly on the phylogeny and partly on the ecology of the species concerned [GARTLAN, 1968; STRUHSAKER, 1969]. The social nature of primates and their tendency to form social groups are features firmly rooted in phylogeny. When they are prevented from forming social bonds, serious and long-term effects on individual adjustment and social behavior are observed [HARLOW and HARLOW, 1962].

Manipulation of the habitat, restriction of space and alteration of feeding patterns, all of which occur when animals are captured and confined, can have far-reaching effects on the form and frequency of previously established social patterns [GARTLAN, 1968]. Such changes may eventually be accompanied by physiological deterioration, making them irreversible. The problem is to establish the phylogenetic norms for each species and to demonstrate how frequently and how profoundly they are modified, attenuated or even suppressed. Many of these questions can be investigated by observation and experiment in the laboratory as well as in the field.

Variability of social structure within primate genera and families is genetically determined. Some genera, such as *Cercocebus*, are socially conservative and appear to live exclusively in multi-male groups. Others, such as *Cercopithecus*, show considerable specific variation in their social structures. Now that a considerable number of field studies of wild primates have been made, it has become apparent that even within single species there may be significant variations in social structure. Some of these variations are obscure in function, others are apparently related to the immediate features of the environment.

Fig. 1. Adult female *Erythrocebus patas* and infants drinking at waterhole, Waza, arid sahel savanna.

At present, however, it seems that neither phylogenetic affinity (such as membership of the same genus) nor ecological similarity (such as inhabiting the same forest) is necessarily of value in predicting the basic features of social structure such as group size, adult sex ratio, or the types and frequencies of social interaction. It is the purpose of this paper to document and discuss the significance of these variations by reference to African primates.

African primates live in a rich variety of habitats: arid sahel savanna (fig. 1), montane grassland, woodland savanna (fig. 2), gallery forest, lowland primary rain-forest, secondary growth forest (fig. 3), and swamp and mangrove forest. Savanna and open-country species are geographically more widespread than those living in the rain-forest. *Papio anubis* of the savannas extends from about 7° W to 40° E; *Papio leucophaeus* of the rain-forest does not even span 5° of longitude. *Cercopithecus aethiops tantalus* has been studied both in Waza, Cameroon (14° 35′ E) and in Murchison Falls, Uganda (33° E). On the other hand, *Cercopithecus lhoesti preussi*, a species of the Cameroon

Fig. 2. Adult male *Cercopithecus aethiops* alarm-barking from top of *Canarium schwein-furthii* tree, woodland savanna.

montane rainforest, is distributed within an area of less than 75×75 miles. The widespread geographical distribution of savanna species increases the probability of their encountering different or variable environmental conditions. As a result they provide more material for discussion in this paper than do forest species.

The Range of Variability in Social Structure

Cercopithecus nictitans in Cameroon and *Cercopithecus mitis* in Uganda are congeners inhabiting similar biotopes of high forest, and they seem to have similar social structures. Both species form groups with a single adult male and a mean of 14 individuals [ALDRICH-BLAKE, 1970].

Living in the same forests as *C. nictitans* in Cameroon and often combining with them into multi-species associations is *Cercopithecus mona*. This species forms groups of up to 38, which may include five adult males, and in

Fig.3. Secondary growth forest in Cameroon. Habitat of *Cercopithecus mona.*

this respect it differs markedly from the sympatric *C.nictitans* and behaves more like its savanna congener, *C.aethiops.*

C.aethiops lives in semi-arid sahel savanna through a range of inter-mediate habitats extending as far as the secondary forests adjacent to lowland rain-forest. Here it may be found in multi-species associations including such typically rain-forest species as *C.nictitans* and *C.mona.* In Cameroon *C. aethiops* ranges from Waza in the savanna (11° 25′ N, 14° 35′ E) with an annual rainfall of 65 cm to near the village of Kak in the forest (4° 52′ N, 9° 42′ E) where the annual rainfall is 269 cm. The size of the social groups found in *C.aethiops* is variable, ranging from fewer than 10 to over 50 indi-viduals. Group size appears to be related in some way to the environment for mean group size tends to be inversely correlated with the mean annual rainfall (table I).

STRUHSAKER [1969] maintained that heterosexual groups with only one adult male were typical of most *Cercopithecus* species. He also believed that all-male groups were not characteristic of forest primates. Recent data tend to

Table I. Relationship of group size in *Cercopithecus aethiops* and annual rainfall in five African locations

Location	Mean annual rainfall cm	Habitat type	Mean group size (n in brackets)
Amboseli (2° 40′ S, 37° 10′ E)	38	savanna	24[1]
Waza (11° 25′ N, 14° 35′ E)	65	sahel savanna	23 (2)
Chobi (2° 12′ N, 32° 12′ E)	114	woodland savanna	18 (3)
Lolui (0° 07′ S, 33° 40′ E)	159	forest-savanna	12 (46)
Kak (4° 52′ N, 9° 42′ E)	269	secondary forest	8 (3)

1 'Typical' group given in STRUHSAKER [1967a]; no n given.

indicate that these views may be an oversimplification, and that forest *Cercopithecus* species may vary considerably. In Cameroon I frequently found both *C. mona* and *C. erythrotis* in large, multi-male groups. In the Ivory Coast, a group of *Cercopithecus campbelli lowei* has been recorded with one male of more than five years of age, two males of four years, two of three and two of two years [BOURLIÈRE *et al.*, 1969]. Although large all-male groups have not been observed in forest *Cercopithecus*, pairs of adult males were seen not infrequently.

Adult males of *C. aethiops* often change from one heterosexual group to another. A male may leave one group and remain solitary for a few days or a few weeks, whilst gradually attaching himself to another group [GARTLAN, 1966]. Single adult males of one species of *Cercopithecus* also join groups of other species, and may remain with them for several weeks. An adult male *C. erythrotis* was observed to join and move as part of a *C. nictitans* group for a period of two and one-half months. Similar observations have been made on a male of *Cercopithecus pogonias* which also moved with a *C. nictitans* group for three weeks. *C. aethiops* males also often join with *Papio* groups in the savannas. Such behavior is probably typical of the entire genus *Cercopithecus*.

Thus, whilst the phylogenetic affinities of a species may on occasion help us to predict its behavior patterns, on many occasions they are of no value for this purpose. Ecological affinity between species may also be of very limited predictive value. In the Waza savanna, both *C. aethiops* and *Erythrocebus patas* are sympatric; but the former are found as usual in multi-male units, wheras the latter live in their characteristic one-male groups [STRUHSAKER and Gartlan, 1970]. Durham [1971] working with *Ateles paniscus* in Peru, noted that the mean group size of 18.5 at an altitude of 275 m decreased sharply at 567 m to 11. Counts of *Cercopithecus lhoesti preussi*, primarily a montane species in Cameroon, showed a range in group size from two to eight animals. However, there was no significant difference between counts obtained at Barombi Mbo (275 m) and those from Bambuko (1100 m).

These data indicate the wide range of variability in primate social structure and the lack of any single causal factor for it. It is proposed now to examine factors important in the formation of specific social structures. These are considered to be of three distinct but interrelated types: phylogenetic factors, density-dependent factors and environmental factors.

Phylogenetic Factors

Phylogenetic differences important in social organization have been clearly demonstrated within the genus *Macaca* by Rosenblum *et al.* [1964]. Groups of *Macaca nemestrina* and *Macaca radiata* were kept in identical cages; and it was shown that while *M. radiata* normally huddled together, *M. nemestrina* individuals tended to sit separately.

The patas monkey, *E. patas*, occurs in one-male heterosexual groups and all-male groups [Hall, 1965]. One of the features which appears to be important in the formation of all-male groups is the attraction which the brightly colored scrotal-anal area holds for other males, infant, juvenile, and subadult. Very often these animals come to touch, handle, and sniff the ano-genital area of adult males, even if the latter are from another group. No female was ever seen to show this interest [Gartlan and Gartlan, in press]. The formation of patas all-male groups appears, therefore, to be permitted by the possession of a characteristic, highly visible, ano-genital area. It would be surprising, then, if similar behavior patterns occurred, for example, in the forest-dwelling *C. nictitans*, which has a small and cryptically-colored scrotum.

Bearing in mind the above example, one might expect to find differences between the social structures of drills and mandrills, in which the adult males have brightly-colored circumanal skin, and those of savanna baboons of the *Papio* genus, which are not so colored. Some of the differences were listed in Gartlan [1970]. Similarly, the fact that adult male drills possess a secretory sternal gland which is used to mark trees might also lead to the prediction that this species is more territorial than its savanna congeners, a prediction that is still being investigated.

Not all phylogenetic constraints on social structure are based on obvious anatomical differences. There is, for example, no obvious physical reason to account for the different social distances of *M. nemestrina* and *M. radiata*. Behavior patterns may be ecologically inflexible, not based on any clear physical structures, and at the same time be highly stable.

Recent studies of *Theropithecus gelada* [R. I. H. DUNBAR, personal communication] have shown that the geladas have a social system based on the one-male-group type of social structure, which is effected through a rather different set of behaviors than that of *Papio hamadryas*. The subadult male gelada leaves his all-male group and joins a harem as a 'follower' [in the sense used for *P. hamadryas* by KUMMER, 1968]. There he cultivates the young juvenile females and builds up associations with them. A few years later he leaves, presumably taking with him the now adult females to set up an independent unit. But the two harems subsequently remain in very close contact, and together they now constitute the so-called 'clan'. Any harem seems free to wander off at will, although one rarely does for more than a few days.

Many of these processes are similar to those known for *P. hamadryas*, but the behavioral means through which they operate are very different. The geladas are very docile compared with the hamadryas, and the younger males may start to cultivate the attention of juvenile females within the harem without attracting the aggressive attentions of the adult harem male. The subadult male hamadryas, on the other hand, has to start by kidnaping babies and infants. Similarly, among geladas harem males do not solicit new females. It is the female that comes and chooses a harem to join; wheras the hamadryas females are captured by males, whether they like it or not. The differences are instructive, since the two species have arrived at a similar system through totally different pathways.

The evolution of similar physical characteristics in closely-related genera may serve entirely different functions in each. Both *E. patas* and *C. aethiops* adult males have a striking red-white-and-blue pattern: the blue of the scrotum, the white fur in the medial line between anus and scrotum, and the red perianal skin. However, in *C. aethiops* these features are used in an aggressive display [red-white-and-blue display, STRUHSAKER, 1967b] consolidating a linear social hierarchy; whereas in *E. patas* they serve to attract other males and seem to be functional in the formation of all-male groups [GARTLAN and GARTLAN, in press].

Distinct species-specific differences in behavior can develop over periods of time that are relatively short in evolutionary terms. *Erythrocebus* is considered to have diverged from *Cercopithecus* relatively recently [DANDELOT,

1968]; but the frequency of sexual behavior of the two species is nevertheless very different. Copulation in *E.patas* was never observed by me at Waza, and HALL [1965] saw it only seven times during his study in Uganda. This total may be surpassed in a single *C. aethiops* group during a single morning at any time of year. *C. aethiops* in captivity copulates throughout the year, and even pregnant females are apparently sexually receptive. Copulation in *E.patas* is markedly seasonal in captivity [ROWELL, personal communication].

In general, all the foregoing behavior patterns show features indicating that they are under direct genetic control, and that they are probably not the result of learning. Thus *P. hamadryas* and *P. anubis* hybrid males behave towards their females in a way predictable from the number of genes of each species in their make up [NAGEL, 1971]. Nevertheless, it is possible for a genetically controlled and stable feature, such as the loud calls of *Cercopithecus* males, to be suppressed if the appropriate releasing stimuli do not occur. In zoos in Europe, for example, these loud calls are rarely heard; for the males are never out of visual contact with their group members, and there are usually no other conspecific groups in the vicinity.

Density-Dependent Factors

I carried out studies of *C. aethiops* on Lolui Island, Lake Victoria, Uganda (0° 07′ S, 33° 40′ E), and at Chobi in the Murchison Falls National Park, Uganda (2° 12′ N, 32° 12′ E). The ecology of these two study sites was markedly different. At Chobi the habitat was disturbed and deteriorating owing to overgrazing by elephant, *Loxodonta africana* and hippopotamus, *Hippopotamus amphibius*. Lolui was a rich and regenerating habitat [JACKSON and GARTLAN, 1965]. The population density of *C. aethiops* at Chobi was approximately 57 per square mile; whereas on Lolui it was about 225 per square mile. Day ranges at Chobi were often three, four, or more times longer; and the groups were more often more dispersed[1] than was usual on Lolui.

Associated with the greater population density per square mile on Lolui was a smaller territory size. Indeed, the territories there were so small that it was impossible for group members to be out of earshot of one another, and unusual for them to be out of visual contact. Unlike Amboseli, where a 'core'

---

1   The position of at least 75% of the group was plotted on a map and the smallest possible circle or ellipse was drawn around the group. The dispersal index is the diameter or, in the case of the ellipse, the mean of the major and minor axes.

area within the home range was defended [STRUHSAKER, 1967a], the home range boundaries and the territorial ones on Lolui coincided closely. The mean territorial area of 20 groups on Lolui was 0.06 square miles, and that of one group was 0.03 square miles. At Chobi where there was no territorial behavior, home ranges were 10 to 15 times larger than on Lolui. These distribution differences were reflected in differences in social behavior. On Lolui, with its very dense population, a form of scent-marking with the chest by adult males was observed [GARTLAN and BRAIN, 1968]. This has never been observed in any other wild population of *C. aethiops*, although it has been seen occasionally in captivity. It appears to be density dependent.

Studies made in captivity have shown that drastic changes in normal social structure and behavior can occur when population densities become high [GARTLAN, 1968]. It is not clear to what extent these changes are reversible in the early stages; but there is ample documentation that when groups have been kept in such conditions for a long period of time, physical changes and physiological deterioration set in and are probably irreversible in the majority of cases. Overcrowding to this extent is rarely encountered in the wild, but it is apparent that such features of social organization as the presence or absence of territorial behavior in *C. aethiops* are density-dependent.

Primates living in low-productive habitats may be widely dispersed, as was seen at Chobi. This scattering in turn limits the opportunities for social interaction, and for the learning of relationships between individuals. Inter-group grooming was observed at Chobi on five occasions; no territorial behavior, such as the chasing-out of intruding groups, was observed. This situation was in direct contrast to Lolui, where inter-group grooming was never seen and a group that intruded onto another's territory was immediately and noisily chased out.

Differences were also observed in intra-group social encounters; again they probably reflect the degree of crowding. Intra-group social grooming was common on Lolui. During 19 days of observation selected at random from the whole study, in one group of 14 there were 0.68 incidents per hour; whereas for the main group at Chobi over an equivalent period, there were only 0.26 incidents per hour in a group of 26 animals. The frequency of aggressive incidents within the same social groups over the same periods showed a similar pattern. On Lolui there was an average of 0.28 incidents per hour; whereas at Chobi there were only 0.18 incidents per hour.

When one compares the two areas, so different in ecology and density of population, it is apparent that at heavily-populated Lolui the groups were more integrated, the social groups were more discrete, and the individuals within the groups interacted more frequently than at sparsely populated Chobi.

Transient Environmental Factors

The influence of environment on social structure is seen in a variety of behavior from group dispersal to the frequency of vocalization. These behaviors are the most labile of the three categories discussed and are readily modified by environmental changes.

GARTLAN [1966] demonstrated that seasonal factors, particularly the onset of the rains on Lolui, affected the type and availability of food for the C. aethiops population, conditions that, in turn, affected the dispersion of the social groups. At the onset of the rains, when there was a flush of edible grass shoots, the groups were significantly more compact than at other times of the year.

There is evidence that in some species of primate the dorsal carrying of infants may be determined by environmental factors. Infant drills and mandrills have not been recorded riding dorsally in the wild, and it would clearly be impractical for them to do so in the dense thickets of undergrowth where they live. (It is possible, however, that this restriction is a phylogenetic one.) ROWELL [personal communication] records that in a forest-living population of P. anubis, the mothers carried their infants dorsally, as is typical in this species by the time the infants are a few weeks of age. However, when these animals were studied in cages, the dorsal carrying pattern failed to appear.

Loud calls made by adult males are a common feature of the behavior of forest Cercopithecus. These calls are presumed to impart information about certain features of the environment to other members of the population. A given call may function as what MARLER [1961] called an 'identifier', 'designator', 'prescriptor' or 'appraiser'; several functions may be present in the same call. Thus in C. mona, the typical adult male loud call begins with one to three paired booms, followed by a series of loud hacks [STRUHSAKER, 1970]. The booms, which do not carry very far in the forest, function in promoting group cohesion. The hacks carry for considerable distances, and have a group-spacing function. Environmental and social influences on Cercopithecus loud calls are of two main types. Either the form and the content of the message may vary, or the vocal pattern may be completely suppressed. Single adult males in the forest are usually completely silent, giving neither contact, alarm or loud calls. Vocal patterns are not therefore spontaneous reactions to stimuli, but rather reactions within a specific social context. Similar behavior is observed in areas where primates are hunted heavily. All alarm calls tend to be suppressed, even within social groups, and the animals tend to move off silently.

C. aethiops lacks a loud call, but this is not a necessary feature of savanna life. E. patas possesses one; it was described by HALL [1965]. In Uganda it was rarely heard; but at Waza, where the population density of patas was higher, it was common. It was made by the adult male of a heterosexual group when either a single adult male, an all-male group or another heterosexual group approached. This situation suggests that the call is directly homologous with the loud calls of forest Cercopithecus, for both patas and forest Cercopithecus are quite often out of visual contact with other group members.

This two-phase roar (whoo-wherr) of E. patas was never made by any adult male in all-male groups, despite the fact that these appeared in many ways to be close and discrete units [GARTLAN and GARTLAN, in press], having, in an attenuated form, many of the features of the social structure of heterosexual groups. There is no doubt that these males were physically capable of making the call, for they were often fully mature and in the prime of life. It can only be concluded that the call was being suppressed in much the same way that single adult males of forest Cercopithecus species suppress their calls. The all-male group of patas is apparently not an appropriate social context for the call.

Summary and Conclusions

Social structure and organization are not unitary and have evolved along many lines. Social organization is composite and susceptible to many features of the environment.

The fact that there are similarities in the number of adult males found in the social groups of primates of different species has previously led to the conclusion that the social organization of, for example, E. patas, P. hamadryas, T. gelada, and C. nictitans is similar. In reality they are very different. In this paper I have tried to demonstrate that although both phylogenetic affinities and ecological similarities may be important in the expression of social organization, the precise, limiting factors must be explored in each case.

This paper has discussed some of the features of social organization that are susceptible to environmental factors. I have classified the most important features into three categories. The first category contains fixed, genetically controlled factors which often depend on the presence of specialized structures. Examples are the red-white-and-blue displays of *C. aethiops* and *E. patas* and the presence of sternal glands in *Papio leucophaeus*. Factors that are also genetically determined but do not affect external structures include the differences in individual distance between *M. nemestrina* and *M. radiata* and in the behavior of adult males towards females in *P. hamadryas* and *T. gelada*. These factors are relatively inflexible. They are probably minimally susceptible to learning or to environmental influence, although they can drop out of a repertoire.

The second category includes density-dependent factors. Crowding, for example, leads to the development of social hierarchies; and if the situation is allowed to become chronic, physical and physiological changes occur that render the hierarchy a relatively stable and inflexible feature of the social organization of a particular group. Because of this property, factors in this second category have often been mistaken for phylogenetic factors. Territoriality is another feature of social organization that is often density-dependent. Further, in habitats where productivity is low, animals are more often more dispersed than in more productive areas. When animals are more dispersed, the amount of social grooming and of intra-group aggression decreases. Groups become less cohesive and less structured, having fewer opportunities to establish formal, learned relationships.

The third category includes those transient factors which depend on a particular or a changing environment. For example, the distribution of food may be affected by seasonal factors, which may then affect the distribution of the social group in its home range. Another example is seen in the calls of forest *Cercopithecus* adult males. These calls are stable characters, but their number and type are directly related to specific environmental factors such as population density and the frequency of predation. The calls are assumed to provide information about environmental features to other groups and other members of the same group. Thus, a species living in an area where predation is high and population density is low will have a very different vocal repertoire from another population living in an area of high population density but low predation. This third category therefore includes the most labile characters.

Social structure thus represents the interaction of several kinds of factors of different degrees of flexibility and sensitivity to the environment. Factors in one category may mimic factors in another and cause confusion. The complete analysis of social structure therefore requires the identification of the components and a description of how, in a particular case, they interact to produce the observed patterns of behavior.

Acknowledgements

The field work reported here from Cameroon was supported by MRC grant No. G969/524/B. I wish to thank Mr. ROBIN DUNBAR for information on *Theropithecus gelada* and Drs. J. E. WRIGHT and B. O. L. DUKE for criticism of this manuscript.

References

ALDRICH-BLAKE, F. P. G.: Problems of social structure in forest monkeys; in CROOK Social behaviour in birds and mammals, pp. 79–101 (Academic Press, London 1970).

BOURLIÈRE, F.; BERTRAND, M., and HUNKELER, C.: L'écologie de la mone de lowe *(Cercopithecus campbelli lowei)* en Côte d'Ivoire. La terre et la vie *2:* 135–163 (1969).

DANDELOT, P.: Preliminary identification manual for African mammals. Primates: Anthropoidea (Smithsonian Institution, Washington 1968).

DURHAM, N. M.: Effects of altitude differences on group organization of wild black spider monkeys *(Ateles paniscus)*. Proc. 3rd int. Congr. Primatol., vol. 3, pp. 32–40 (Karger, Basel 1971).

GARTLAN, J. S.: Ecology and behavior of the vervet monkey *(Cercopithecus aethiops pygerythrus)*, Lolui Island, Lake Victoria, Uganda; Ph. D. thesis, University of Bristol (1966).

GARTLAN, J. S.: Structure and function in primate society. Folia primat. *8:* 89–120 (1968).

GARTLAN, J. S.: Preliminary notes on the ecology and behavior of the drill, *Mandrillus leucophaeus* Ritgen, 1824; in NAPIER and NAPIER Old world monkeys: evolution, systematics and behavior, pp. 445–480 (Academic Press, New York 1970).

GARTLAN, J. S. and BRAIN, C. K.: Ecology and social variability in *Cercopithecus aethiops* and *C. mitis*; in JAY Primates: studies in adaptation and variability, pp. 253–292 (Holt, Rinehart & Winston, New York 1968).

GARTLAN, J. S. and GARTLAN, S. C.: Quelques observations sur les groupes exclusivement mâles chez *Erythrocebus patas*. Ann. Fac. Sci. Yaoundé (in press).

HALL, K. R. L.: Behaviour and ecology of the wild patas monkey *Erythrocebus patas* in Uganda. J. Zool. *148:* 15–87 (1965).

HARLOW, H. F. and HARLOW, M. K.: Social deprivation in monkeys. Sci. Amer. *207:* 137–146 (1962).

JACKSON, G. and GARTLAN, J. S.: The flora and fauna of Lolui Island, Lake Victoria. A study of vegetation, men and monkeys. J. Ecol. *53:* 573–597 (1965).

KUMMER, H.: Social organization of hamadryas baboons. Bibl. primat., vol. 6, pp. 1–189 (Karger, Basel 1968).

MARLER, P.: The logical analysis of animal communication. J. theoret. Biol. *1:* 295–317 (1961).

NAGEL, U.: Social organization in a baboon hybrid zone. Proc. 3rd int. Congr. Primatol., vol. 3, pp. 48–57 (Karger, Basel 1971).

ROSENBLUM, L. A.; KAUFMAN, I. C., and STYNES, A. J.: Individual distance in two species of macaque. Anim. Behav. *12:* 338–342 (1964).

STRUHSAKER, T. T.: Ecology of vervet monkeys *(Cercopithecus aethiops)* in the Masai-Amboseli game reserve, Kenya. Ecology *48:* 891–904 (1967a).

STRUHSAKER, T.T.: Behavior of vervet monkeys *(Cercopithecus aethiops)*. Univ. Calif. Publ. Zool. *82:* 1–64 (1967b).

STRUHSAKER, T.T.: Correlates of ecology and social organization among African cercopithecines. Folia primat. *11:* 80–118 (1969).

STRUHSAKER, T.T.: Phylogenetic implications of some vocalizations of *Cercopithecus* monkeys; in NAPIER and NAPIER Old world monkeys: evolution, systematics and behavior, pp. 365–444 (Academic Press, New York 1970).

STRUHSAKER, T.T. and GARTLAN, J.S.: Observations on the behaviour and ecology of the patas monkey *(Erythrocebus patas)* in the Waza Reserve. Cameroon J. Zool. *161:* 49–63 (1970).

Author's address: Dr. J. S. GARTLAN, Institut de Recherches médicales, BP 55, *Kumba, South-West Province* (République Fédérale du Cameroun, West Africa)

Symp. IVth Int. Congr. Primat., vol. 1: Precultural Primate Behavior,
pp. 102–123 (Karger, Basel 1973)

Play and Art in Apes and Monkeys[1]

B. RENSCH

Zoologisches Institut der Universität, Münster (Westf.)

A. Introduction

All mammals (particularly the young ones), several species of birds, and possibly also some species of fish show a behavior which we call playing because it resembles very much some simpler types of play of human children. This parallelism already indicates that playing probably belongs to proto-cultural behavior, the roots of which have been developed very early in vertebrate phylogeny. Playing of apes and monkeys can also be used to investigate whether and how far these animals already have some basic aesthetic feelings. An analysis of such rudimentary predispositions is of interest with regard to the origin of human art. A wealth of corresponding observations and experiments can now be used to judge the meaning of play and possible aesthetic feelings.

B. Play

1. Development, Characteristics, and Meaning

Playing has been observed in most species of mammals and several species of birds, particularly parrots, ravens, crows, and jackdaws. Whether similar behavior of some fish can be regarded as playing is an open question [BEACH, 1945]. At least *Mormyridae*, a family with a relatively large cerebellum,

---

1   This article is dedicated with the author's cordial wishes to his friend, the great Indian ornithologist, Dr. SALIM ALI, on the occasion of his 75th birthday.

seem to play with little twigs, which they try to balance on their foreheads [MEYER-HOLZAPFEL, 1960]. One can perhaps object that we are dealing here with vacuum activity of a food-searching or nest-building instinct. Without regard to such doubtful cases we must assume, however, that playing originated relatively early in the phylogeny of vertebrates.

Adult mammals mainly play in relaxed situations, sometimes together with their offspring, or in zoological gardens, where they do not have to be concerned about food or enemies. In young mammals, playing is often released by a surplus of vigor. Solitary play can also be caused by exploratory or curiosity behavior, which is developed to a particularly high degree in monkeys and apes. New objects arouse attention [BERLYNE, 1960], and this often induces the animals to touch them and to play with them if they prove to be movable. WELKER [1956 a, b] and MENZEL et al. [1961] observed that chimpanzees always spent more time in contact with new objects than with well-known ones. Generally, young animals showed a more intense exploratory behavior than adult ones. MENZEL et al. could also prove that new objects were touched for a particularly long period when they had an unusual shape or size [cf. also MENZEL, 1963]. Free-living Japanese monkeys (Macaca fuscata), however, relatively seldom manipulated new objects; normally, caution prevailed [MENZEL, 1966].

In order to get an idea of the phylogenetic development of curiosity behavior, my former student, WÜNSCHMANN [1963], compared fish (Cyprinus and Carassius), birds (quails and jackdaws), and an ape (our young chimpanzee, 'Julia'). Although only these few species could be tested, the results of long series of trials clearly indicated that the attraction to new stimuli increased markedly from lower to higher levels of the vertebrates and that the ape showed by far the highest degree of curiosity. By another method, GLICKMAN and SROGES [1966] tested how 200 animals in a zoological garden reacted to objects which were unknown to them. Generally, primates and carnivores showed much more curiosity than rodents and species of other more primitive orders of mammals.

Playing is normally released by a *playing-mood* that is often characterized by certain gestures. This situation is particularly true when an animal invites a partner to join in play, but it sometimes also occurs when two animals begin to play with a movable object. Normally, an animal gives a push against a possible playmate or an object, for instance a ball, or makes a mock attack. Chimpanzees and also some monkeys show a typical play-face which resembles human smiling: the corners of the mouth are a little drawn up so that only the teeth of the lower jaw become visible.

Some types of play, particularly those which use the same motor patterns employed in hunting and fighting, can be released by very generalized stimuli. Very different species may, therefore, play with one another, for example, different species of monkeys and apes. In the garden of the ornithologist Lord Medway near Kuala Lumpur, Malaysia, we observed the play of a tame gibbon and a dog; these animals had played together in the same way many times before. Although the gibbon fought mainly with arms and hands and the dog with the mouth, they always reacted well to one another. All apes and monkeys also like to play with men.

The *general characteristics* of human and animal play have been a subject of speculation for many years. Plato thought that playing could be regarded as exercising later adult behavior. The German philosopher, Immanuel Kant, stressed the accompanying pleasure and the lack of a consummatory action; our German poet and historian, FRIEDRICH SCHILLER [1795] discussed the release of a surplus of energy and a certain degree of 'freedom'; SPENCER [1880] emphasized the exercising of all possible actions and the release of certain reactions by dummies. The first detailed analysis of animal play was, however, published by GROOS [1896/1930, 1898]. In the first edition of his book, he over-emphasized the exercising of later adult behavior, but in the third German edition [1930], he discussed nearly all characteristics which we recognize today to be important.

The following characteristics are those we now recognize [cf. ALLEMAN, 1951, cited by MEYER-HOLZAPFEL, 1956b; MEYER-HOLZAPFEL, 1956a; LOIZOS, 1967]: the occurrence of components of normal instinctive behavior in incomplete and variable sequence lacking consummative actions; repetition of single components; tendency to exaggerate certain movements; social inhibitions, particularly avoidance of injuring the partner; possible use of inanimate objects or individuals of other species as substitute playmates; possible interruption in every stage by stronger stimuli (loud noise, appearance of enemies, need of defecation, and so on); transmission of playing mood to other individuals, particularly to playmates; a certain degree of freedom in inventing new individual or experimental play, sometimes leading to new nervous and muscular coordinations [EIBL-EIBESFELDT, 1951]; and correspondence to simple human play. Normally, playing occurs only in a relaxed situation, free from physiological pressures and threats from the environment.

There can be no doubt that the play of young animals has a clear *selection value*. If this were not the case, play would not have originated independently among many orders of higher animals in a species-specific manner and would not always have been maintained. As young animals spend much time in playing, they exercise their later behavior and their nervous and muscular co-ordinations. They learn the extent of their physical capacities; monkeys, for instance, learn how far their arms can reach and how far they are able to jump. In addition, they learn how to react to different situations in their environment. Playing also enhances social contact, a fact which is very important for monkeys and apes. CARPENTER [1965], who worked on howler monkeys, and SOUTHWICK et al. [1965], who worked with rhesus monkeys, stressed this effect. Moreover, playing has the advantage of canalizing strong drives, particularly aggressive drives, into a harmless outlet [GROOS, 1930].

2. Types of Primate Play

Within the scope of a single paper it will not be possible to review the vast and rather scattered literature on primate play. I must restrict myself to discussing some characteristic examples. Compared with the play of other mammals, the play of primates is less stereotyped and more interspersed with individual 'free' actions. The specific manner of playing is determined by the structure of the body and by inborn instincts and types of behavior. Young gorillas, for instance, normally play by running about on the ground and wrestling; whereas young orangutans, being more adapted to arboreal life, like to clasp a branch or another object with one of the four extremities while playing. Spider monkeys and capuchine monkeys often use their fifth ex-tremity, the tail, for clasping things in play. It is impressive to see how much the play of monkeys and apes resembles the simpler play forms of human children. In order to emphasize the protocultural meaning of play, I shall always refer to similar types of human play.

It is not always possible to distinguish the normal movements of young primates from true play when the animals only touch objects or other indi-viduals or when they are rolling on the ground. However, somersaults, which they often repeat several times, may be regarded as play. BIERENS DE HAAN [1952] observed somersaults in a young chimpanzee; HARRISON [1964], in a young orangutan; and SCHALLER [1965], in free-living young gorillas. When our young chimpanzee, Julia, first had an opportunity to run about on a lawn at the age of two and a half years, she apparently found the experience highly pleasurable. She showed a typical play face and made a series of somersaults spontaneously. Several other types of *solitary play* have been observed in apes, in captivity as well as in the wild. SCHALLER [1963, 1965] noted that young gorillas liked to slide down oblique tree trunks. REYNOLDS observed young

Fig. 1. Young gorilla playfully sliding on his forearms on the smooth floor [after RENSCH and DÜCKER, 1966].

chimpanzees repeatedly climbing up a liana and sliding down it [REYNOLDS and REYNOLDS, 1965; REYNOLDS, 1966]. Also, FOSSEY [1970] reported young mountain gorillas sliding down playfully from trunks, slopes, or the backs of their mothers. A young orangutan about two years old and a lowland gorilla of similar age (with which we were working at the Yerkes Regional Primate Research Center in Atlanta) liked to slide on their forearms on the smooth floor of the room (fig. 1) [RENSCH and DÜCKER, 1966]. SOUTHWICK *et al.* [1965] observed rhesus monkeys jumping playfully on the backs of cattle and then jumping from one animal to the other. The monkeys also pulled on the tails of the cattle and hit them on the nose.

The spontaneous play of human children is very similar. Children like to roll on the ground, slide down slopes, swing on branches, turn somersaults, and scamper about in the same manner as young apes and monkeys do.

Fighting play is very similar in all primates and is also similar to playful fighting of carnivores. The inhibition against injuring the partner, particularly a biting-inhibition, is a common characteristic. In most cases, a mock aggression against a partner or one of the parents serves as an invitation to play. A young gorilla drums with both hands on his chest, belly, or upper thigh, on the ground, on the wall, or on the legs of his human partner. Young orangutans invite their playmates only by touching and pulling them. Then they try to clasp, wrestle, and bite, in a manner similar to gorillas, chimpanzees, capuchine monkeys, or mongooses [RENSCH and DÜCKER, 1966; cf. also SCHALLER, 1963].

The play behavior of human children in mock-fighting situations is very similar to that of monkeys; children, like monkeys, often use teasing as an invitation to play. GROOS [1930] remarked that such teasing may be regarded as an early stage of human comic behavior. Children chase and catch one another and wrestle and avoid injuring their playmates. However, the fighting drive is also canalized by human culture into more complicated types of play with rules to be learned, such as football, tennis, and other games. Some of these are purely intellectual, for example, chess, cards, and so on.

Our chimpanzee, Julia, and our capuchine monkey, Pablo, liked to play tug-of-war with me when I held a flap or a cord in my hand and they could catch the other end. SCHALLER [1963] observed young gorillas playing together in this manner with the stalk of a plant. The instinctive root of this type of play, which is also common among human children (and dogs or mongooses), is probably the defense of food.

Components of sexual behavior, such as the mounting during play [observed in rhesus macaques by SOUTHWICK *et al.*, 1965], do not appear in all species. But a female monkey or ape often handles a human or animal doll by pressing it against her chest like a real baby. In the Budongo Forest, REYNOLDS observed a young chimpanzee who played 'mother and child' with a still younger ape by embracing and kissing him and allowing him to climb on her back or to hang on her belly [REYNOLDS and REYNOLDS, 1965; REYNOLDS, 1966]. In our culture, girls like to play 'mother and child' because they are given a doll as soon as they can handle it. Girls from more primitive cultures do not play very much in this manner.

Individually developed 'experimental' play is particularly interesting because it indicates a certain mental versatility. Such types of play are often initiated by curiosity behavior that leads monkeys and apes to touch all kinds of objects and to try to find out if they are movable. KÖHLER [1921] emphasized that chimpanzees often try to discover all the possible functions of an object. They use sticks to beat, poke, dig, or climb on. More complicated play can also be influenced by experience. Therefore, even planned play can be developed. KÖHLER [1921] mentioned a typical example. A chimpanzee lured chickens to the lattice fence of an enclosure by feeding them, and then suddenly pricked them with a stick. Human children also try out the various possible functions of sticks and other objects.

In order to avoid repeating well-known examples of such individual play, I will only report some observations which we made in our Institute in Münster. Our capuchine monkey, Pablo, once discovered that the handle of a broken tennis racket could be used as a 'jimmy'. He took the handle in his tail, climbed

Fig. 2. Capuchine monkey, Pablo, playfully tries to break up the food container from a mechanically favorable position.

up to his sitting board, and put the thinner part of the handle between sitting board and wooden food-container. Then he climbed up a little on the lattice, pressed vigorously on the other end of the handle (that is, on the longer lever arm), and from this favorable position tried to break up the food-container. He repeated this play several times (fig. 2; first observation by Dr. A. Nolte).

Once we placed a stool in the large cage of our chimpanzee, Julia. After some time, the ape suddenly 'leap-frogged' over the stool as a boy does and repeated this new play several times. She also liked to cover her head with a cloth (fig. 3) and would grope about without seeing where she went. This behavior, which was observed several times, very much resembled the playing of blindman's buff of human children. VAN LAWICK-GOODALL [1971] made a similar observation in the wild. A chimpanzee had pulled a stolen cover over his head and tried to poke after his fellows. Even a mongoose *(Herpestes edwardsi)* repeatedly played 'blindman's buff' by slipping his head and neck into a narrow paper bag and running about blindly [RENSCH and DÜCKER, 1959].

Fig. 3. Chimpanzee, Julia, covers her head with a cloth in order to play blindman's buff.

After Julia had learned to use human instruments in our experiments, she used them in new situations when playing. She had learned, for example, to open a box with a large screwdriver to get food. Later on she tried to loosen other screws in her cage. The manipulatory interest of chimpanzees also leads them to imitate the use of other human objects. HAYES [1952], for instance, noted that her famous young chimpanzee, Viki, imitated human actions while playing with a nail file, a powder box, a lipstick, and the ear-piece of a telephone.

Playful 'self-decoration' of apes also has a particular meaning with regard to protocultural behavior. KÖHLER [1921] noted that his chimpanzees liked to put pieces of cloth, tendrils of plants, twigs, or cords around their necks. As the apes often showed a certain accentuation of their gait when draping themselves in this way, KÖHLER interpreted such behavior as a pleasant enhancement of their self-esteem, but not as a real 'decoration'. HAYES [1952] mentioned similar play in her chimpanzee, Viki. Our capuchine monkey, Pablo, also liked to put a piece of cloth or a towel around his neck. SCHALLER [1963, 1965] observed free-living gorillas who put cushions of moss on their heads or lobelias on the napes of their necks. Little human children also like such self-decoration. Although this behavior of apes probably has nothing to do with aesthetic pleasure, it may have been important in our ape-like ancestors for the development of true adorning. Besides, in *Homo sapiens*, self-decoration means both adornment and enhancement of self-esteem.

Summing up, we may state that the *play of monkeys, apes, and human children have the same physiological base and that many types of play in animal and human children are identical or very similar in form*. Man, however, also developed more voluntary and planned types of play which are transmitted by tradition from generation to generation.

Playing was also important in the development of human culture, because it was combined with exploratory behavior and could lead to new types of action. This process occurred particularly after the general functions of an object, a stick, a stalk, or a stone had been discovered [cf. also RENSCH and DÖHL, 1967; KORTLANDT, 1968]. Playing contributed to the development of the 'precision grip' from the 'power grip' [NAPIER, 1962]. Thus the *Homo ludens* or probably the *Homo erectus ludens* became the *Homo faber*.

C. Art

Darwin's theory of sexual selection was based on the assumption that animals are able to perceive 'beauty'; for instance, female birds are capable of judging the magnificent plumage or the nice song of the male birds. WALLACE [1889] opposed this explanation and pointed out that females reacted by inborn instinct to the stimuli of male characteristics. Many experiments have confirmed WALLACE's opinion. Investigations on bird songs and paintings of apes, however, allow us to assume that higher animals may nevertheless have some basic aesthetic feelings like those of man. Of course, in animals only the formal components of sensations can be taken into consideration. In order to make a judgment about their feelings, it is necessary first to state what kinds of components are aesthetically effective in man.

Since FECHNER's first investigations on preference of colors and shapes [1876], many relevant analyses have been carried out. We now know that the following characteristics are especially preferred by young human children [cf. ZIEHEN, 1923, 1925]. These characteristics can also contribute to the aesthetic feelings of monkeys and apes. (1) Saturated colors are preferred to unsaturated ones; (2) primary colors are preferred to mixed colors; (3) brilliant colors on the surfaces of objects are preferred to non-brilliant ones; (4)

Fig.4. Capuchine monkey, Pablo, choosing between regular and irregular patterns [after RENSCH, 1957].

rhythmical repetition of equal components pleases because it facilitates comprehension [ZIEHEN, 1923, 1925] and produces 'pleasure of recurrency'; (5) in the same fashion, bilateral and radial symmetry also produce aesthetic pleasure; (6) steady curves, like circles, spirals, wave-lines, and so on are preferred to irregular curves; (7) conspicuous lines or shapes are preferred to indistinct ones; (8) a certain balance of excitation between the left and right half of a picture is more pleasant than an unbalanced arrangement; (9) when two objects of different colors have to be combined, the same colors or conspicuously different colors are preferred to *nearly* equal colors (principle of disappointed expectation). Of course, adult humans often react in a different manner because they are accustomed to certain colors and shapes.

In order to draw conclusions regarding positive feelings of monkeys and apes, one can offer them a choice of play objects which differ only in these visual components [RENSCH, 1958, 1965, 1969]. This method allows one to make quantitative statements about the preference for particular components. However, it is first necessary to test, by different combinations, whether an animal prefers a color or a pattern because it corresponds to an innate releaser. For instance, a green monkey *(Cercopithecus aethiops)* that I tested preferred white to blue, red, and yellow in significant percentages. White very probably corresponded to an inborn releaser, as this species shows an area of white skin around the eyes which becomes more visible when the monkey becomes

Table I. Preference for rhythmical and symmetrical patterns to irregular patterns by a capuchine monkey *(Cebus apella)* and a green monkey *(Cercopithecus aethiops)*. Six patterns (3 rhythmical or symmetrical and 3 irregular) were offered in each trial. The first choice has always been counted. Significant values marked by +. No preferences were noted for an irregular pattern [after RENSCH, 1957]

Pairs of pattern	Capuchine monkey		Green monkey	
	n	% of choices of regular pattern	n	% of choices of regular pattern
	93	65.7 +	144	62.8 +
	220	61.4 +	254	52.7
	299	58.8 +	213	53.2
	256	60.2 +	295	59.1 +
	190	56.4	140	52.2
	225	57.3	195	55.2
	201	56.7	140	55.0
	100	66.0 +	404	57.0 +

positively or negatively excited. A preference or avoidance of white would, therefore, not necessarily mean positive or negative aesthetic feelings. Our capuchine monkey, Pablo, preferred yellow to other colors.

Preference determined by inborn releasers is much less probable when one uses *black and white geometrical patterns*. I offered Pablo squares of cardboard arranged in a checkerboard manner; half of the pieces showed rhythmical patterns, and the other half showed irregular patterns of similar black and white content (fig.4). I always noted which type of pattern the monkey took first. Since all the pieces of cardboard could be used as playing objects, Pablo, of course, chose both types; he did not always inspect them carefully. It was, therefore, necessary to make a great many trials in ever-changing arrangements, in order to assess his preference. Trials in 1957 proved that the monkey preferred rhythmical and symmetrical patterns to more irregular ones (table I). This preference corresponded to the aesthetic preferences of man. When the monkey had to choose between three concentric squares and three concentric circles, he preferred the latter pattern. Perhaps the steady flexure of the circles determined this preference.

A green monkey was not so attentive to the differences of the patterns. Although she preferred three of the same patterns as Pablo, in all other cases she chose at random; but in no case did she prefer the irregular pattern (2,706

Fig. 5. Capuchine monkey, Pablo, combines colored cubes [after RENSCH, 1957].

single trials). When I offered three types of white cardboard of equal surface but of different shape (circular or irregularly rounded), the capuchine monkey (76 choices) and the chimpanzee (93 choices) preferred the circular ones in significant percentages; and the green monkey (100 choices) did so in a nearly significant percentage (44%; normal probability 33.4%).

Our capuchine monkey and a young female chimpanzee also learned to *pin together two colored cubes* (fig. 5). Sometimes they combined three or four of them. In 8,004 trials with the capuchine monkey and 324 with the chimpanzee, the capuchine monkey normally preferred red, yellow, green, or blue to grey, a pattern that corresponds to the behavior of human children. The capuchine monkey and the chimpanzee preferred, in significant percentages, combining cubes of the same color. In one series of trials, I offered six grey and six blue cubes, and the capuchine monkey combined grey with grey in 69.5% of the 95 cases; random combination would only yield 20%. When he got six yellow and six grey cubes, he combined yellow with yellow in 51.5% of 186 trials (that is, in more than double the normal probability); and when he got six green and six grey cubes, he combined green with green in 53.2% of 109 trials. When six black and six grey cubes were offered, he combined black with black in 53.8% of 208 cases, more than 2½ times as often as could be expected if normal probability prevailed.

The young chimpanzee showed similar behavior. She was allowed to pick up the cubes from the floor and put them in a box or in my sleeve after experiments with 12 cubes of two colors were finished. In several trials the ape collected five or all six cubes of one color first. For example, in ten cases, five or six grey ones were collected first and then the six (or seven) red ones; but in two cases, five red ones were collected first. This preference for collecting the same color, similar to matching from sample, can perhaps be explained by the easier comprehension of action and 'complexibility' and thus can produce 'pleasure of recurrency' in a manner similar to the preference of rhythmical patterns.

A special series of trials yielded an interesting result. I offered the capuchine monkey 7 different colors gradually shading over from yellow to dark red. Each color was represented by six cubes. As the monkey could in each case choose two of these 42 cubes, 28 different combinations were possible. The normal chance of combining two cubes of the same color was only 1.74% and of any two different colors, 4.18%. The main results of 2,490 trials are shown in table II. In significant percentages, the monkey combined yellow 1 with yellow 1, yellow 2 with yellow 2, red 6 with red 6 and red 7 with red 7. These results mean that he had two kinds of preference: (1) He liked to choose the clearer colors, yellow 1 and red 7, and to a lesser degree, yellow 2 and red 6. (2) So far as clear colors were concerned, he combined the same color in percentages significantly higher than normal probability; the percentages of his combinations of the clearer colors (1 + 2 and 6 + 7) were also significantly higher than normal probability. The percentages of combinations of the intermediate neighboring colors were much too small, and the differences from normal probability were also significant. Human behavior is similar. We like genuine sameness and clear contrast more than a combination of very near shades; *apparent* sameness seems to produce negative feelings. We are dealing here with the 'principle of disappointed expectation'.

The results of all these investigations show that we may suppose positive feelings in certain monkeys and apes in those instances when they indicate preference for rhythmical and symmetrical patterns to more irregular ones and for the combination of the same or clearly different colors to very similar ones. It is, therefore, justifiable to ask whether the *scribbling and painting* of apes and capuchine monkeys also indicate certain basic aesthetic feelings. Drawings and paintings of these species have repeatedly been induced and analyzed, first by Garner in 1900 [cited by Margoshes, 1966], Shepherd [1915], Kohts [1935],

Table II. Percentages of combinations of cubes of 7 different colors shading from yellow (1) to dark red (7) chosen by a capuchine monkey. Each shade is represented by 6 cubes (the subject could, therefore, always choose 2 from a total of 42 cubes). The combined data are taken from 5 series of trials totalling 2,490 combinations. Percentages differing significantly in positive or negative direction from normal probability are marked by + and – (unimportant combinations are omitted) [after RENSCH, 1957]

Combinations of colors	Normal probability in %	Produced combinations in %
A. Same color		
1 + 1	1.74	7.20 +
2 + 2	1.74	3.19 +
3 + 3	1.74	1.30
4 + 4	1.74	0.89 –
5 + 5	1.74	0.41 –
6 + 6	1.74	3.97 +
7 + 7	1.74	4.85 +
B. Very different colors		
1 + 7	4.18	8.82 +
2 + 7	4.18	7.98 +
1 + 6	4.18	4.08
C. Neighboring shades		
1 + 2	4.18	7.93 +
2 + 3	4.18	2.78 –
3 + 4	4.18	2.01 –
4 + 5	4.18	1.30 –
5 + 6	4.18	2.13 –
6 + 7	4.18	7.19 +

and SOKOLOWSKY [1928], and later by KLÜVER [1933], SCHILLER [1951], HAYES [1952], MORRIS [1958a, b, 1961, 1962], RENSCH [1958, 1965, 1969], KORTLANDT [1959], GOJA [1959], and others. MORRIS has discussed his intensive investigations in detail in his 1962 book. It will, therefore, not be necessary to treat these questions exhaustively here.

When an ape gets a pencil and a sheet of paper and the experimenter presses the hand of the animal a little on the paper so that the ape sees that he can make a mark, he will normally begin to scribble at once. The same holds true for capuchine monkeys. This scribbling then becomes a kind of play, which the animals like to repeat several times. These productions can be very different, depending upon the size and the format of the sheet and upon its vertical, inclined, or horizontal position. Further differences depend upon whether the pencil is pointed or broad and whether the animal paints with a smaller or broader brush. The bearing of the body also has some influence. Some apes were sitting on a chair, others were crouching; our capuchine monkey was sometimes standing.

There are also large differences between species and between individuals of the same species. Some animals do not pay attention to what they are producing and simply make bundles of more or less parallel oblique lines by pulling the pencil or brush to and fro. Such productions strongly resemble the scribblings of human children of one and a half years of age [cf. Eng, 1931]. They are only the result of the pleasant play of making marks, and aesthetic feelings surely had no influence.

In other cases, the scribblings and paintings are probably guided by primitive aesthetic feelings. The chimpanzee, Jenny, for instance, began in the center of the sheet and filled the paper rather harmoniously with short crabbed lines [Goja, 1959]. The scribblings of Kohts' [1935] and Schiller's [1951] chimpanzees were similar [cf. also the corresponding figures in Morris, 1962]. Sometimes our chimpanzee, Julia, and also a male chimpanzee, Fips, of the Zoological Garden in Münster, tried to cover the sheet of paper uniformly with black color. Klüver [1933] observed that a capuchin monkey painted the floor of his cage almost uniformly with red color. This tendency to produce a uniformly colored surface may possibly be caused by positive 'proto-aesthetic' feelings.

The assumption that basic aesthetic feelings partly guide painting by apes is, however, much more problematic in many other cases. Chimpanzees often produce fan-like bundles of lines showing a certain rhythm and dynamics and a balance of the left and right half of the painting. In man, the sight of such paintings produces basic aesthetic pleasure. But it is doubtful whether this is also the case in apes. The fan-like arrangement can be caused by drawing the lines to the place where the chimpanzee is sitting. Similar types of drawings have been produced by our capuchine monkey, Pablo, when he began to draw with chalk on the wall of his cage in 1954. On vertical sheets of paper fixed on the wall, Pablo later painted more curved lines, which were not arranged in a fan-like manner (fig. 6).

It is easier to analyze paintings when a chimpanzee successively uses brushes with different colors, because then one can better judge the sequence of lines. In many of these cases it becomes clear that the rhythm of lines and the

Fig. 6. Painting with black color on a vertical sheet (31 × 35 cm) by capuchine monkey, Pablo [after RENSCH, 1961].

balance of the left and right half of the picture comes about in a planned manner. Sometimes Julia first made a fan-like picture with wide gaps between the lines and then filled the gaps with lines of the next color (fig. 7). A picture by the chimpanzee, Congo, which Dr. DESMOND MORRIS kindly gave me, shows that the ape first made a fan-like picture covering a large part of the sheet. With the next color he began not so far to the left side and ended not so far to the right side. And a new color was then applied still more toward the middle, so that in all stages of development a balance of the left and right half of the picture was maintained.

MORRIS [1962] could prove this tendency also experimentally. He offered his chimpanzee a rectangular sheet showing a grey square or a black-bordered square on the left or right side. Such figures attract the attention of apes and capuchine monkeys and they normally draw lines over the given figure. SCHILLER [1951] investigated this tendency by using different figures. MORRIS found the same reaction. His chimpanzee, Congo, also often drew over the given figures; but in several cases he drew only on the opposite side of the sheet. This must not yet be explained as the ape's intention to balance the drawing. It is more probable that the large empty space induced the ape to draw there. When MORRIS, however, displaced the squares only a little from the center, Congo scribbled also on the empty space near the center. These cases seem to indicate a real sense of balance.

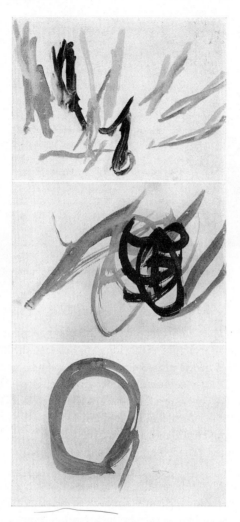

Fig. 7. Three pictures by chimpanzee, Julia, late stage; successively painted (a) with red, yellow and black; (b) with red, yellow, and blue; (c) with red.

When one observes the actions of apes or capuchine monkeys during scribbling or painting, it becomes clear that a normal tendency exists to *begin in the middle* of the sheet. I could prove this tendency best by offering in succession to a female chimpanzee a pencil and 14 circular sheets of paper (diameter 14.5 cm). When she began to scribble, I interrupted her actions after two or three seconds, so that only a few lines could be drawn. In 11 cases it

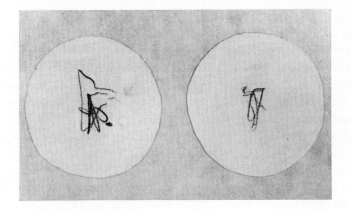

Fig.8. Centralized scribblings with pencil by chimpanzee, Lotte. Diameter of sheets: 7.5 cm [after RENSCH, 1961].

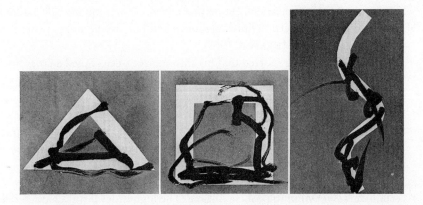

Fig.9. Chimpanzees, Lotte and Julia, tried to paint along white patterns on grey background [after RENSCH, 1961].

became clear that the ape had begun in the middle of the sheet and in four she started in the *exact* center of the sheet (fig. 8). Trials with two other chimpanzees and with smaller sheets of paper (diameter 7.5 cm) yielded similar results.

The *format of the sheet* also influences the direction of the lines. Narrow, rectangular sheets normally become filled with lines which are adapted to the longer side of the paper, the lines being sometimes nearly parallel to the long borders. Circular sheets induce more curved lines (fig. 9). Sometimes, but by no means always, a circle drawn by the experimenter caused curved lines to be drawn; and two thick parallel lines induced lines of similar direction. Such adaptations of the direction of lines became more conspicuous when I fastened white strips of paper onto grey cardboard. The strips formed triangles, squares, spirals, S-shaped, and zig-zag figures. In most cases the two chimpanzees, Lotte and Julia, when I confronted them with this task, tried to draw their lines on these white strips (fig. 9).

It was never possible, however, to induce true *copying*. Together with my former assistant, Dr. Döhl, I tried to do so in the following way. With white chalk we drew a triangle on a blackboard, and below the triangle we drew a square. Beside the triangle we indicated three corners by marking three points and by the side of the square, four points, and indicated with a finger that Julia should connect these points. She soon learned to do this and drew a not very exact triangle and square. Then we made the points successively finer. But as soon as the points became nearly invisible, the chimpanzee made only more or less circular lines which resembled neither a triangle or a square. She had only learned to produce a certain drawing movement, but she did not pay attention to the figures close by [RENSCH, 1965].

SHEPHERD [1915] reported that his chimpanzee once copied a cross and a 'W', although in an imperfect manner. I guess that his ape may perhaps also have learned corresponding movements of the hand without trying to copy. Perhaps more similar to true copying was the behavior of a young gorilla in the Zoological Garden of London. HUXLEY [1942] and FORD observed that the ape four times traced with his forefinger along the contour of his shadow, which was clearly represented on the white wall.

MORRIS [1962] noted that paintings by the chimpanzee, Congo, showed a certain *development* in the course of time. The highest level was characterized by more or less circular figures, inside which the ape painted some lines and twirls. Our Julia showed the same development. She began with bundles of oblique lines and fan-like paintings; later on she also produced more circular pictures (fig. 7), and in some cases she made fan-like paintings in which she filled the gaps more carefully.

Summing up all investigations about color and pattern preference and scribblings and paintings of apes and capuchine monkeys, we can assume that the actions were guided to some degree by positive feelings which correspond to basic aesthetic feelings of man. The following main findings certainly speak in favor of this assumption: (1) preference of rhythmical and symmetrical patterns and of steadily curved lines to more irregular patterns; (2) preference for combining cubes of equal or conspicuously different colors and avoiding combinations of nearly equal colors; (3) centralization of scribblings and paintings; (4) adaptation of the direction of lines to the format of the sheet or to figures marked on the paper by the experimenter; (5) balance of left and right half of a painting.

However, the strength of these tendencies should not be overestimated. Sometimes apes do not notice figures which the experimenter has drawn beforehand on a sheet, or they produce inharmonious scribblings, or they paint over a picture and disturb its primary harmony. It is also improbable that apes have an idea of what they will draw. On the other hand, the paintings of apes which arouse a certain aesthetic pleasure in man are not just pictures selected from a collection but are often produced in series. A certain aesthetic quality is also proved by the fact that several exhibitions of ape paintings in different countries have been arranged by societies of modern art. Experts in art who did not know that a painting was not a human production were often almost enthusiastic about the dynamics, the rhythm, and the harmony of the pictures.

Human art at the level of Neanderthal men already shows a high degree of perfection. It is, therefore, probable that we have to search for the root of these abilities in man's very early ancestors. Basic aesthetic feelings must be phylogenetically very old, because birds also show clear preferences for rhythmical and symmetrical patterns [RENSCH, 1958] and for acoustical harmony. We feel justified in regarding the expression of proto-aesthetic feelings in animals, as well as animal playing, as typical protocultural behavior.

References

BEACH, F. A.: Current concepts of play in animals. Amer. Naturalist *79:* 523–541 (1945).

BERLYNE, D. E.: Conflict, arousal, and curiosity (McGraw Hill, New York 1960).

BIERENS DE HAAN, J. A.: Das Spiel eines jungen solitären Schimpansen. Behaviour *4:* 144–156 (1952).

CARPENTER, C. R.: The howlers of Barro Colorado Island; in DEVORE Primate behavior, pp. 250–291 (Holt, Rinehart & Winston, New York 1965).

EIBL-EIBESFELDT, I.: Beobachtungen zur Fortpflanzungsbiologie und Jugendentwicklung des Eichhörnchens *(Sciurus vulgaris).* Z. Tierpsychol. *8:* 370–400 (1951).

ENG, H.: The psychology of children's drawings (Routledge & Kegan, London 1931).

FECHNER, G. T.: Vorschule der Aesthetick; 2nd ed., vol. 1 (Breitkopf & Härtel, Leipzig 1897).

FOSSEY, D.: Making friends with mountain gorillas. Nat. Geographic *137:* 48–67 (1970).

GLICKMAN, S. E. and SROGES, R. W.: Curiosity in zoo animals. Behaviour *26:* 151–188 (1966).

GOJA, H.: Zeichenversuche mit Menschenaffen. Z. Tierpsychol. *16:* 369–373 (1959).

GROOS, K.: Das Spiel der Tiere; 1st and 3rd ed. (Fischer, Jena 1896/1930).

GROOS, K.: The play of animals; E. L. BALDWIN, transl. (Appleton, New York 1898).

HARRISON, B.: Kinder des Urwaldes. Meine Arbeit mit Orangutans auf Borneo (Brockhaus, Wiesbaden 1964).

HAYES, C.: The ape in our house (Gollancz, London 1952).

HUXLEY, J.S.: Origins of human graphic art. Nature, Lond. *149:* 637 (1942).

KLÜVER, H.: Behavior mechanisms in monkeys (University of Chicago Press, Chicago 1933).

KÖHLER, W.: Intelligenzprüfungen an Menschenaffen (Springer, Berlin 1921).

KOHTS, N.: Infant ape and human child. Scient. Mem. Museum Darwinianum (1935).

KORTLANDT, A.: Tussen Mens en Dier (Wolters, Groningen 1959).

KORTLANDT, A.: Handgebrauch bei freilebenden Schimpansen; in RENSCH Handgebrauch und Verständigung bei Affen und Frühmenschen, pp. 59–102 (Huber, Bern 1968).

LAWICK-GOODALL, J. VAN: In the shadow of man (Collins, London 1971).

LOIZOS, C.: Play behaviour in higher primates: a review; in MORRIS Primate ethology, pp. 176–218 (Weidenfeld & Nicolson, London 1967).

MARGOSHES, A.: The first chimpanzee drawing. J. genet. Psychol. *108:* 55–57 (1966).

MENZEL, E.W.: The effects of cumulative experience on responses to novel objects in young isolation-reared chimpanzees. Behaviour *21:* 1–12 (1963).

MENZEL, E.W.: Responsiveness to objects in free-ranging Japanese monkeys. Behaviour *26:* 130–150 (1966).

MENZEL, E.W.; DAVENPORT, R.K., and ROGERS, C.M.: Some aspects of behavior toward novelty in young chimpanzees. J. comp. physiol. Psychol. *54:* 16–19 (1961).

MEYER-HOLZAPFEL, M.: Über die Bereitschaft zu Spiel und Instinkthandlungen. Z. Tierpsychol. *13:* 442–462 (1956a).

MEYER-HOLZAPFEL, M.: Das Spiel bei Säugetieren; in HELMCKE und VON LENGERKEN Handbuch der Zoologie, vol. 8/2, pp. 1–36 (de Gruyter, Berlin 1956b).

MEYER-HOLZAPFEL, M.: Über das Spiel bei Fischen, insbesondere beim Tapirrüsselfisch *(Mormyrus Kannume* Forskål). Zool. Garten, N.F. *25:* 189–202 (1960).

MORRIS, D.: The behaviour of higher primates in captivity. Proc. 15th Int. Congr. Zool., pp. 94–98 (London 1958a).

MORRIS, D.: Pictures by chimpanzees. New Scientist *4:* 609–611 (1958b).

MORRIS, D.: Primate aesthetics. Natur. Hist. *70:* 22–29 (1961).

MORRIS, D.: The biology of art (Methuen, London; Knopf, New York 1962).

NAPIER, J.R.: The evolution of the hand. Sci. Amer. *207:* 56–62 (1962).

RENSCH, B.: Ästhetische Faktoren bei Farb- und Formbevorzugungen von Affen Z. Tierpsychol. *14:* 71–99 (1957).

RENSCH, B.: Die Wirksamkeit ästhetischer Faktoren bei Wirbeltieren. Z. Tierpsychol. *15:* 447–461 (1958).

RENSCH, B.: Malversuche mit Affen. Z. Tierpsychol. *18:* 347–364 (1961).

RENSCH, B.: Über ästhetische Faktoren im Erleben höherer Tiere. (Naturwiss. Med.) *2:* 43–56 (1965).

RENSCH, B.: Ästhetische Grundprinzipien bei Mensch und Tier; in ALTNER Kreatur Mensch, pp. 134–144 (Moos, München 1969).

RENSCH, B. und DÖHL, J.: Spontanes Öffnen verschiedener Kistenverschlüsse durch einen Schimpansen. Z. Tierpsychol. *24:* 476–489 (1967).

RENSCH, B. und DÜCKER, G.: Die Spiele von Mungo und Ichneumon. Behaviour *14:* 185–213 (1959).

RENSCH, B. und DÜCKER, G.: Manipulierfähigkeit eines jungen Orang-Utans und eines jungen Gorillas. Mit Anmerkungen über das Spielverhalten. Z. Tierpsychol. *23:* 874–892 (1966).

REYNOLDS, V.: Budongo (Natural History Press, New York 1966).

REYNOLDS, V. and REYNOLDS, F.: Chimpanzees of the Budongo Forest; in DEVORE Primate behavior, pp. 368–424 (Holt, Rinehart & Winston, New York 1965).

SCHALLER, G. B.: The mountain gorilla (University of Chicago Press, Chicago 1963).

SCHALLER, G. B.: The behavior of the mountain gorilla; in DEVORE Primate behavior, pp. 324–367 (Holt, Rinehart & Winston, New York 1965).

SCHILLER, F.: Über die ästhetische Erziehung des Menschen. 27. Brief. HOREN, 1795. Sämtliche Werke, vol. 12, p. 119 (Cotta, Stuttgart).

SCHILLER, P. H.: Figural preferences in the drawings of a chimpanzee. J. comp. physiol. Psychol. *44:* 101–111 (1951).

SHEPHERD, W. T.: Some observations on the intelligence of the chimpanzee. J. anim. Behav. *5:* 391–396 (1915).

SOKOLOWSKY, A.: Erlebnisse mit wilden Tieren (Haberland, Leipzig 1928).

SOUTHWICK, C. H.; BEG, M. A., and SIDDIQI, M. R.: Rhesus monkeys in North India; in DEVORE Primate behavior, pp. 111–196 (Holt, Rinehart & Winston, New York 1965).

SPENCER, H.: Principles of psychology; 3rd ed., vol. 2 (London 1880).

WALLACE, A. R.: Darwinism (Macmillan, London 1889).

WELKER, W. I.: Some determinants of play and exploration in chimpanzees. J. comp. physiol. Psychol. *49:* 84–89 (1956a).

WELKER, W. I.: Effects of age and experience on play and exploration of young chimpanzees. J. comp. physiol. Psychol. *49:* 223–226 (1956b).

WÜNSCHMANN, A.: Quantitative Untersuchungen zum Neugierverhalten von Wirbeltieren. Z. Tierpsychol. *20:* 80–101 (1963).

ZIEHEN, T.: Vorlesungen über Ästhetik, vol. 1 and 2 (Niemeyer, Halle 1923 and 1925).

Author's address: Prof. Dr. BERNHARD RENSCH, Zoologisches Institut der Universität, Badestrasse 9, *D-44 Münster (Westf.)* (FRG)

Symp. IVth Int. Congr. Primat., vol. 1: Precultural Primate Behavior,
pp. 124–143 (Karger, Basel 1973)

Pongid Capacity for Language Acquisition

An Evaluation of Recent Studies

G. W. HEWES

Department of Anthropology, University of Colorado, Boulder

My interest in animal capacities for language goes back almost exactly
half a century. In 1922, I proposed a study of 'bird language', which was to be
based on a program of carefully listening to bird sounds. When it was pointed
out to me that this would not necessarily tell me what they said, I added, 'Well,
I'll watch and see what they do after every sound.' Although this conversation
was written down at the time by my mother, nothing came of the plan, since I
was only five years old. The next step in my involvement with this topic came
21 years ago, in 1951, when I wrote to Keith J. Hayes, who at that time was still
engaged in studying the home-raised chimpanzee Viki. The Hayes, it turned
out, had already considered the possibility of teaching Viki sign language. The
notion was not new. Almost 50 years ago, ROBERT M. YERKES [1927, p. 180],
one of the great pioneers in primate research, wrote:

'I am inclined to conclude from the various evidences that the great apes have plenty
to talk about, but no gift for the use of sounds to represent individual, as contrasted to
racial, feelings or ideas. Perhaps they can be taught to use their fingers, somewhat as does
the deaf and dumb person, and helped to acquire a simple, nonvocal "sign language".'

And, more than 250 years before YERKES' statement, we find in the *Diary*
of SAMUEL PEPYS the following entry for August 24, 1661:

'By and by we are called to Sir W. Battens to see the strange creature that Captain
Jones hath brought with him from Guiny; it is a great baboon, but so much like a man in
most things, that (though they say there is a species of them) yet I cannot believe but that
it is a monster got of a man and a she-baboon. I do believe it already understands much
English; and I am of the mind it might be taught to speak or make signs.'

PEPYS, for a layman, was of a scientific turn of mind; in 1684 he was
elected President of the Royal Society. However, he let the matter of teaching
a chimpanzee (for such it seems to have been rather than a baboon) drop; the

same evening he went to the theater to see Thomas Betterton play in *Hamlet, Prince of Denmark*.

In my letter to the HAYES of July 3, 1951, I stated that 'a nonvocal approach to teaching her (Viki) symbolic communication (should) be given a fair trial'. I went on to suggest that this might take the form of the sign language used by deaf-mutes or the somewhat similar sign language of the Indians of the North American Plains. I specifically referred to the more or less iconic sign languages, not to finger-spelling. The proposed training, I continued, would require that Viki be exposed 'at least several hours a day, for many months at a minimum, to human companions who not only use conventional hand-arm gestures in dealing with her alone, but in conversing with each other'. If at the end of six months or a year of this, she still showed no inclination to make such signs on her own account, and to employ them meaningfully to effect changes in her environment through social interaction on the level of social communication, we might be ready to look beyond the nerves and muscles of her mouth and pharynx for the reason that apes do not have language. 'Assistants', I added, 'could probably be obtained from among the deaf-mute population of this country.' Keith Hayes, in his reply, said that Viki already managed to express most of her wants quite effectively and that, while the idea of sign-language training had been considered, there were other research problems of greater interest. In any case, even if Viki had then been introduced to sign language, the outcome would probably have been disappointing; she died not long after our exchange of letters.

It was not until the late 1960's that R. A. GARDNER and B. T. GARDNER [1969], at the University of Nevada in Reno, actually undertook such an experiment with the chimpanzee Washoe, with results at least somewhat familiar to most of my readers. I should explain that their experiment was entirely independent in conception, although they report that, not long after it began, they learned from Hayes that someone had proposed such an experiment to him years earlier.

In the 1950's, the Zeitgeist was unfavorable to any such research, whether in linguistics, psychology, or anthropology. Linguists, basing their ideas about behavior on the theories of a distinguished array of S-R psychologists, outdid the most orthodox Watsonians in their opposition to what they called 'mentalism'. SKINNER'S *Verbal Behavior* [1957] was probably regarded by the majority of American linguists at that time as an excellent statement of what underlies the human capacity to use language.

Linguists had been inhibited for decades (by a tradition going back to 1866, when the 'Société de Linguistique de Paris' banned papers dealing with language origins), from open speculation about how language began. To be sure, the ban had been based on the mediocre or crack-pot character of most of the glottogenic theories being offered at the time; but there were unstated anxieties also produced by the precocious application of Darwinism to linguistics. The Paris taboo was not wholly effective, and German writers exhibited an enthusiasm for language-origin theorizing that no ukase of French scholars could suppress. Darwin had of course addressed the problem in *The Descent of Man* and approached it also in his treatise on the expression of the emotions in man and animals.

French scholarly misgivings also did not deter that quixotic pioneer of primate field studies, RICHARD LYNCH GARNER. His pathetic effort, with its born-too-soon employment of primitive phonographic cylinders, was an explicit attempt to decode the 'language' of the apes and monkeys in their natural habitat. Despite his glowing reports, subsequent investigators could find no evidence either that apes could speak or that they had languages of their own. Furness was able to elicit only a few poorly enunciated words from his orang. The Hayes could do no better with Viki.

There the matter might have rested, had it not been for the unexpected widening of fossil discoveries of the Australopithecines, which I think have played a more important role in all this than is generally realized. Dart's 1924 announcement of *Australopithecus africanus* had an inauspicious reception. Only after Robert Broom and others began to accumulate specimen after specimen of these creatures of the early Pleistocene and late Pliocene was the anthropological community forced to include these forms in the scenario of hominization. Thanks to these Australopithecine remains and the cultural remains associated with them, there took place the development of a new and more sensitive style in hominization model-building. Speculations about language origins took on a new respectability. The same period also brought forth modern primatology, which I date from the early 1950's, in Africa and Japan.

In view of this history, it is worth noting that neither GARDNER and GARDNER [1969, 1970, 1971] nor PREMACK [1971a, b, c], who worked with the now famous Sarah, consciously developed their chimpanzee experiments in the light of paleoanthropology. To the contrary, the roots of their studies appear to lie in the contradictions within neo-behavioristic psychology. Both the GARDNERS and PREMACK have left it to others to compare their respective experiments and (up to now at least) to evaluate the bearing their work may have upon glottogenic theorizing. I have been impressed by the fact that the fossil Australopithecines had brains not much, if at all, larger than existing pongids, although they had achieved regular bipedalism and a modicum of tool-using beyond that observed in modern anthropoid apes. It seems reasonable to me that Australopithecine linguistic capabilities were not any further in advance of those of the existing pongids than their tool-making and tool-using behaviors.

I have reviewed elsewhere the implications of this view [HEWES, 1971, 1973]. To reduce these ideas to a very brief statement, I can say that for a long time before the evolutionary emergence of articulate vocal language, the early hominids probably communicated propositionally by means of hand and arm gestures, supplemented both by other nonverbal signs and by primate vocal calls not yet deserving of the term 'speech'. This proposal may seem at first hearing to be a case of creating a theory to fit some startling new experimental facts. However, the idea that articulate speech was preceded by a long phase of gestural language is quite old; by the eighteenth century, this idea had been considerably elaborated by Condillac and others. In the nineteenth century, the gestural hypothesis for the origin of language was favored by Alfred Russel Wallace and by Edward B. Tylor, one of the major figures in the formative period of scientific anthropology, and in the early twentieth century was further supported by Wilhelm Wundt, a comparably impressive figure in the history of psychology. Yet these worthy scholars lacked acquaintance with the Washoe and Sarah evidence of the GARDNERS and PREMACK and with the fossil data we now possess on the hominization process. Nor did they have access to relatively recent findings in neurology and psycholinguistics, aphasia research, and allied fields that can also be brought to bear on glottogenesis.

Before moving to the principal argument, I must refer to the question of whether the phenomenon of propositional language is other than species-specific to *Homo sapiens*. I think we may by-pass for the moment the question of bee language, however elegant that may be, and also the whole matter of bird-song analysis, even though some features of avian communication may illuminate certain aspects of hominid language emergence. I think it justifiable also to omit consideration of porpoise and dolphin vocal communication, on the ground that nothing to date really supports the idea that they have propositional language, despite their large brains and rich vocal inventories. We are left with the apes and monkeys, and the issue of the extent to which vocal, postural, gestural, and facially expressive demonstrations in these animals may foreshadow human linguistic behavior. My present judgment is that it is just barely possible that field research may yet show that the chimpanzees have somewhat more proto-language than we have been willing to credit them with. JANE H. HILL [1972] has already dismissed this interesting possibility.

Referentially based gestures are exceedingly hard to decode under field conditions; effective decipherment almost has to depend on manipulation. It is perhaps close to science fiction to suggest that we may at some future date be able to send in a participant chimpanzee ethnographer-linguist, trained in a language we can understand, to settle the question. More seriously, I doubt that chimpanzees or other pongids will be found to have propositional language except of the most rudimentary kind. What we know as 'language' has probably arisen only in the hominid lineage as the outcome of the inter-action of several factors in the new life-ways of these animals where hunting and tool-using were only two of the important ingredients.

If gestural communication was the initial pathway to language as we know it, the process could have begun with simple pointing behavior, as out-lined by the Vietnamese scholar TRÂN DÚC THÂO in several papers in *La Pensée* between 1966 and 1970. How elusive such indicative behavior may be is suggested by EMIL MENZEL's [1972] report of ladder-using behavior in a group of young chimpanzees confined in a large outdoor enclosure. The GARDNERS' Washoe experiment, in particular, shows that a pongid cannot only acquire a simple language system, but can coin new terms when the occasion demands. My hunch is that the ability to combine signs and to create new ones is not really 'linguistic', but something comparable to the behavior of apes in carrying out complex behavioral programs, such as termite-ex-traction with tools.

This ability to carry out a program of complex actions that have been discretely learned is a feature of the behavior of so-called higher animals and has been compared more than once to grammatical or syntactical performance in language. As with aphasia, brain injury may disrupt such programs, leading to *apraxia*. Brain-damaged patients may lose the ability to manipulate in any constructive fashion familiar tools and utensils, even though their hands may still be able to carry out innately programmed reflex actions smoothly and skillfully.

ANDREW [cf. discussants' comments, HEWES, 1973] has objected to the gestural model of glottogenesis on the ground that Australopithecines (or whatever fossil form began to use language) might have attained a sufficient degree of vocal control and differentiation, thanks to vocal tract changes presumably stemming from bipedalism, etc., to render a gestural stage of language an unnecessary construct. He argues that *Papio* and *Theropithecus* exhibit an approximation to hominid vocal productivity and that LIEBERMAN *et al.* [1972] have erred in using rhesus macaques and chimpanzees in their comparative study of human achievement of articulate speech. However, I am not much impressed by this contention; the mandibular specializations of the baboons cannot have been paralleled in the line leading to man or modern pongids. Further, if the empirical methods of LIEBERMAN *et al.* are valid, demonstration of the acoustic incapacity of fossil antecedents of modern man to produce articulate speech would seem to be conclusive. Parrots, mynahs, and magpies can produce reasonable facsimiles of human speech; but the bearing of this mimicking on the *origin* of human speech is well known to be peripheral, however remarkable.

HILL [1972] has objected to a gestural derivation of language on the ground that chimpanzees, who spend several hours daily feeding, do not have their hands free to make manual signs. The implication is that by the same logic, a remote human ancestor would have literally had his hands full, with no time to spare for manual communication. Further, she contends, with the rise of tool- and weapon-using, the early hominids would also have had their hands preempted and could not have communicated manually. If nearly continuous feeding restricts the use of the hands, it should also interfere with the use of the voice. The objection seems groundless to me. Chimpanzees, despite their feeding activity, have time enough to engage in a good deal of manual activity unconnected with food-handling; and primitive hunters, encumbered with weapons, somehow manage to communicate gesturally while maintaining silence during the hunt. After all, we have two hands; and even modern people who use gesture extensively manage to get their manual jobs done. It is even possible, as most of us have probably seen in Mediterranean countries, to drive automobiles and trucks while carrying on a gesture-rich conversation with a driving companion or exchanging insults with other passing drivers.

CICOUREL and BOESE [1972a, b], who have worked with deaf children and sign-language communication, make some pertinent points. They have come to believe that both deaf and hearing children possess a basis for a 'primitive sign system', presumably innate, that is utilized even by hearing children after speech has been fully established. In deaf children in institutions where formal sign language is not taught or even allowed, this capacity leads to the emergence of local institutional sign-language traditions, a fact well known to most specialists in the education of the deaf whether they favor or oppose the sign-language method. Further, and this observation is substantiated over and over again by deaf travellers, users of one system of sign language, such as ASL (American Sign Language for the Deaf, the system taught to Washoe), can manage to establish communication with users of other sign-language systems that supposedly lack historical connection with the others. This process seems to require much less time and effort than learning a foreign spoken language.

Thus, CICOUREL and BOESE [1972a, b] describe a socially isolated family of deaf persons in Southern California who had developed an idiosyncratic sign-language system for their own use. None of the family had attended schools for the deaf. The investigators succeeded in decoding this system, although the task proved to be 'a slow one'. My impression is that, if, instead of a sign language, they had been confronted by a totally foreign spoken language such as Tibetan or Hottentot, 'decoding' would not have been merely slow, but practically impossible, unless bilingual interpreters, or at least dictionaries and grammars, could be found. The relatively easier decodability of sign-language systems may reside in (1) their very restricted lexicons, compared with all known natural spoken language systems; (2) their simpler syntactic structure; and (3) the high iconicity of many of the signs, whereas most of the iconicity in speech is obscure, subliminal, or covert, if it exists at all.

Early voyagers and explorers frequently recorded their employment of 'signs' to communicate with peoples whose languages were utterly foreign to them. From these accounts, which I intend to review in another paper, I gather that such communication is initiated almost immediately; and within a few hours or days, fairly complex information is being exchanged, going far beyond the mere expression of basic human survival needs such as food and water. We are repeatedly told in these travel records that by signs 'we were informed that so many days march to the north there is a large lake', or 'two years since, a ship such as yours visited us', or 'if you will give us iron nails, we will send a guide with you who can show you the route to the coast'. Just how this kind of *ad hoc* communication works and on what basic, pan-human abilities it rests has up to now been a topic totally ignored.

If CICOUREL and BOESE [1972a, b] are right in their guess that there may exist a pan-human sign-language propensity, which NORMAN DENZIN [1971] has also reported in young children, we may ask if this may not also exist in *Pan troglodytes* enabling Washoe, Sarah, and several other chimpanzees to acquire quasi-linguistic behavior. We may be quite misled if we continue to insist that this ability has any necessary relationship to vocalization, decoding of sound signals, or other features of what we customarily think of as *language* [WIENER et al., 1972].

Existing sign languages contain, relative to the inventories or lexicons of natural spoken languages, markedly fewer terms. A standard sign-language dictionary has 1,680 English-language glosses. Introductory texts for ASL have around 1,000 terms [RIEKEHOF, 1969]. If these figures seem unusually small, given the fact that most spoken languages have dictionaries containing several thousand words, it is worth noting that a reading vocabulary of about 1,500 Chinese characters or *kanji* is considered adequate for high-school level reading and composition. To be sure, most *kanji*-combinations consist of two characters, which, could theoretically yield 2,250,000 terms, or far more than any known natural language needs. In view of the lexical productivity of paired roots, a proto-language of two or three hundred basic terms might have been sufficient for Late Lower Paleolithic men. This possibility gets us within range of the lexical accomplishments of Washoe and Sarah.

I think we must assume that hominization has been responsible not so much for our being able to manipulate a language system containing a few hundred lexical items, with whatever modest syntax this might also require, but, much more critically, for our being able to master such a system in the virtual absence of formal training procedures. The Washoe and Sarah experiments show us that with highly motivated instruction, chimpanzees can acquire some sort of language. The evidence now accumulating on language acquisition by human children strongly indicates that they can acquire language competence in the teach-yourself situation with surprisingly little explicit adult 'instruction'. The vocal-auditory channel may have something to do with this phenomenon, but just what is not clear. On the basis of the data so far, it seems that chimpanzees would pick up very little ASL or any other feasible language if they were merely exposed to it in the way that most human children are. The idea that parents contribute much to infant language learning seems to be a middle-class delusion.

I shall not try here to account for the transformation of a hypothetical primordial gestural language system into one employing the vocal-auditory mode. I have already mentioned LIEBERMAN *et al.* [1972] and their reconstructions of the vocal capabilities of Neanderthal and other fossil men. Nor shall I proceed to the fascinating topic of how manual-gestural communication made a comeback, beginning in the Upper Paleolithic, drawing upon the vocal language that had come into being but, in the end, far surpassing it in the forms of writing and graphics generally. This is a topic which has received the attention of André Leroi-Gourhan, and of Alexander Marshack.

Last year, when I first contemplated writing this paper, there had been little critical evaluation of the Washoe or Sarah studies. Now several have appeared [cf. PETERS, 1972; HILL, 1972; MATTINGLY, 1972]. Since these reviews are available, I shall confine my remarks to a few broad questions.

The first serious reactions to the Washoe study were from child-language acquisition specialists who compared Washoe's achievements in ASL with the vocabularies and sentence structures produced by normal hearing children of comparable ages. Although this material on normal spoken-language usage of children is scientifically relevant, it is an error to suppose that the acquisition of ASL is directly equivalent to the hearing child's acquisition of speech; we must compare Washoe with a normal human child who has learned to communicate in ASL. In the case of Sarah, the only human counterpart so far has been a severely retarded deaf child with whom PREMACK [1971b] has employed the plastic-token language he taught to Sarah. Congenital deafness does not of course, imply intellectual deficiency *per se;* but the presumably brain-damaged child with whom PREMACK worked cannot be considered functionally equivalent to Sarah, who seems to be a perfectly normal chimpanzee.

Something also must be said about ASL. The sign language used by the American deaf, which has been carefully analyzed by the linguist STOKOE [1970] of Gallaudet College, should not be thought of as a mere translation into manual gestures of English or of any other spoken language. It is true that it was standardized in late eighteenth century France and has traces of French syntax (e. g., adjectives tend to follow rather than precede nouns they modify), but it is nevertheless not a version of French. It can only be dealt with as a language system *sui generis*, one used by from 100,000 to 200,000 people, of whom most are bilingual in English. GARDNER and GARDNER [1969, 1970] deliberately chose ASL rather than some language constructed for the purposes of their experiment in order to meet the objection, reasonably expectable, that Washoe had not acquired a 'real language'. As it turns out, their choice had many other positive implications, not the least, the possibility of comparing Washoe's linguistic progress with that of children, such as John, whose case I shall describe.

It might be supposed that the literature on the education of the deaf would be replete with well-documented cases of how sign language has been acquired by young deaf children. This is not the case. Until quite recently, authorities on the education of the deaf believed that it was reasonable to defer sign-language training until kindergarten or first-grade level, leaving the profoundly deaf child essentially without language during the very years when the foundations of language are being built into normal hearing children. For this and other reasons, we are left with the remarkable fact that only one case, that of the boy John reported in 1972 by Dr. JACK R. OLSON, is available for direct comparison with Washoe. I am sure that some unpublished records exist of the language development of other born-deaf children acquiring sign language, but for the present we must do what we can with the data from John. Ursula Bellugi, of the Salk Institute, has for some time been videotaping the language behavior of deaf children using sign language, however; and more records for comparison with Washoe should be available in the near future.

John was born with a profound hearing disability, with no sound perception above 1,000 Hz. At this writing (August, 1972), he is more than five years old. Fortunately, we have cumulative vocabulary lists for John and Washoe at about two years and at about three and a half years. John's ASL training began late in his first year, as did Washoe's. The living conditions of the boy and the chimpanzee were not radically different, although Washoe had her own private house-trailer parked in the GARDNER's back yard and John almost certainly slept in his parent's house. Washoe's investigators were trained psychologists, obviously strongly motivated to teach her to communicate in ASL but also very solicitous for her general welfare and socialization. John's parents were equally interested in providing him with a language capacity by means of ASL and were certainly committed to his normal socialization as a human being. John's father was a graduate student in psychology; his mother had a BA degree in sociology. Both the GARDNER's household and that of John's parents would, in sociological parlance, be described as 'upper middle class American'. Both the chimpanzee and the boy had comparable access to house, garden, toys, games, automobile rides, trips around town or into the countryside, books, magazines, etc. Washoe's opportunities to watch TV were limited; John watched TV, but to what extent is not stated. Both child and chimpanzee encountered a variety of other persons.

John was regularly exposed to spoken language on the part of his parents and others; since he was not totally deaf, he was fitted with a special hearing aid that, although it did not enable him to understand speech, evidently permitted him to react to some environmental sounds. By the close of OLSON's direct study of the case, John was lip-reading some 34 words, and in addition he could pronounce three: *ball, boy,* and *bounce.* In contrast, from the start of their project, the GARDNERS took extraordinary pains to prevent Washoe from hearing human speech, although she of course could hear other kinds of sounds, including approximations of chimpanzee emotional cries made by the GARDNERS and their assistants. The rationale for this ban on spoken language in Washoe's presence was that if she were to hear conversations, but could neither understand or participate, she might become frustrated or cease to pay sufficient attention to the sign language. Further, if speech had been allowed to accompany ASL communication, critics might complain that Washoe was reacting to sound rather than visual cues, a circumstance that would diminish the rigor of the experiment by bringing in additional uncontrolled variables. Washoe had contact with a few expert ASL users, students from a school for the deaf; but the GARDNERS and their assistants and John's parents were all less than fluent signers, dependent on ASL manuals or dictionaries to some extent.

HILL [1972] finds the domestic setting for the Washoe project, along with the GARDNERS' staff (which she calls 'an entire department of psychology'), somehow objectionable, apparently because such arrangements are clearly not in keeping with the natural environment of chimpanzees. I think this kind of objection is entirely beside the point. For that matter, John's environment might also be criticized as unrepresentative of the great majority of mankind's.

OLSON [1972] did not try to collect information on John's sentences, although by the age of three he was making sign-language combinations of up to three terms. The GARDNERS recorded as much of Washoe's conversation as possible, although their record was not complete since it was often difficult to write down everything just as it occurred. The GARDNERS took 16-mm films of Washoe's signing and other activity and, had they been more amply funded, might have been able to videotape a still more complete record; but it is obvious that in an experiment stretching over several years, with an energetic subject awake about twelve hours every day, anything approaching a total record would have required an immense budget.

A much more serious difference between the records of John and Washoe has to do with the criteria for including a sign as part of the subject's vocabulary. John's reported lexicon was obtained from his parents, who kept rough track of the signs he was supposed to use regularly and consistently; but no quantification was involved. This procedure is the same as that on which most of our knowledge of language development in hearing children rests, which is to say that it is far from precise and could vary in such a way as to underestimate or to exaggerate the linguistic progress of an individual child. The GARDNERS credited no sign to Washoe unless she had used it *spontaneously* at least five times. Further, the sign had to have some relevance to the situation. In fact, a manual sign uttered completely out of context would have little chance of being recognized. By 'spontaneous' is meant that it did not count if Washoe simply repeated a sign just used by someone else conversing with her. It should be noted in this connection that no such problems arise with Sarah's vocabulary, where each new term was systematically introduced, and on which she was trained until she could employ it in practically error-free fashion.

To have learned a sign in ASL is much like learning a word in a spoken language. One must learn to associate the sign, when it is made by another person, with its proper referent; and since a great many ASL signs consist of

not just one finger position, but hand postures and movements plus place-
ments of hands or fingers on other parts of the signer's body, the decoding
process is not always simple. Further, one must also be able to replicate the
sign, in a way that enables others familiar with the system to decode it in turn.
Sarah, on the other hand, had only to pick plastic tokens out of a compartment
in a box and place them on a magnetized board or to observe such tokens
similarly placed by others. No complicated manual operations were required;
and each token was not only of a quite distinctive, fixed shape and size (unlike
manual gestures, which can vary in size and scope), but further identifiable
by color.

In the Sarah study by PREMACK [1971a, b, c], it seems clear that learning
difficulties having to do with particular signs come from the cognitive problem
of relating the sign to some more or less abstract referent. Ability of chimpanzees to distinguish the shapes, sizes, and colors of small objects has been
amply tested, and the formal properties of the tokens can almost be disregarded in considering the intellectual aspects of the language-learning task.
Not so with ASL signs. It is apparent from a later study based on four young
chimpanzees at Norman, Oklahoma, by ROGER FOUTS [1972] that even where
the level of abstractness or concreteness of the referents is approximately
equal, the signs themselves vary greatly in learnability. The sign for *listen* in
ASL is made by simply touching the ear with the extended index finger. Bruno,
a young male chimpanzee, learned this sign to criterion in only two minutes,
and the mean time for all four animals tested was 9.75 minutes. The sign for
look in ASL involves bringing the hand up to the eyes, palm toward the face,
with the index and middle fingers extended in a V sign, and then turning the
hand away from the eyes (palm away from the face) so that the V moves outward, as if along a line of binocular sighting. The shortest time required to
learn this sign was by Cindy, who took 54 minutes to reach criterion, and the
mean for the four animals was nearly four hours (233.5 minutes). Doubtless,
there are subtle conceptual differences between grasping the idea of *listen* and
look, but the motor complexities of the sign for *look* rather than its semantic
character seem to explain most of the variance. In the same experiment,
FOUTS found that the sign for *string* in ASL caused the greatest amount of
trouble for his chimpanzee pupils, who needed a mean of 316 minutes to
master the gesture. A mean of only 28 minutes was registered for learning to
sign *shoe*, although *shoe* and *string* would seem to have about the same
referential difficulty.

Manual gestures, although they can be used propositionally and with
little display of emotion, are, like spoken words, easily provided with ex-

pressive extra-content; the sign for *come* can be a languid invitation or an urgent plea for help, depending on the accompanying facial expression, body movement, rapidity of its execution, and so on. Colored plastic tokens are affectively neutral. It seems reasonable to suppose that the potential kinesic or paralanguage elements in ASL affect its users and learners rather differently than manipulating plastic tokens affected Sarah and her observers, but in what direction is not evident.

PREMACK was careful in designing his plastic token language for Sarah to avoid iconicity in the signs. *Apple*, therefore, was represented by a blue triangle and not, for example, by a round red object with a projecting bit of stem. Even the signs for colors were deliberately colorless; *banana* lacked either color or shape cues, which would have been easy enough to arrange for in plastic. Like other sign-languages using gesture, ASL is partly iconic, a mixed system of some highly representational elements and others with no apparent formal resemblance to their referents. Still other ASL signs were iconic at one time but would not now be recognized because of cultural changes. *Coffee* (one of the signs used by John) is made with hand movements suggesting the operation of an old-fashioned coffee-grinder, which no modern American child or chimpanzee would be apt to recognize. The ASL sign for *woman* once represented a bonnet-string, appropriate in the late eighteenth century when the system was standardized. The factor of iconicity does not in itself guarantee that a sign will be more easily learned, as we have just seen in the cases of *look* and *string*, although obvious inconsistency of a sign and some outstanding attribute of the referent might cause trouble, as, for example, if the sign for the numeral *two* were to be the display of three fingers (fortunately a hypothetical case).

Sarah, unlike John and Washoe, did not live in a house or house-trailer, but in a laboratory building caged enclosure, in which objects other than those being used in the particular phase of the study were minimal. Her experience was very limited; and she went on no rides and saw no dogs, cats, or sheep, or even book and magazine illustrations. She was, in fact, kept rather strictly confined to the experimental program; and her vocabulary was no larger than it needed to be for the grammatical exercises designed for her by DAVID PREMACK. In this much more limited world, however, Sarah learned to express certain relationships in much more sophisticated fashion than either John or Washoe, an accomplishment for which we are certainly grateful, since the conceptual deficiencies' in the linguistic production of Washoe are nearly offset by the syntactical virtuosity of Sarah. Sarah's seemingly more robot-like performance and lack of creativity is likewise offset by the flexibility and spontaneity of Washoe, who bathed and dried her doll with a towel, learned to sew with needle and thread, avidly looked at pictures in magazines, and otherwise behaved in very complex fashion.

Although I have already made the point that, strictly speaking, Washoe's language performance must be compared to that of a human child of about the same age also trained to communicate in ASL, rather than in spoken English, I shall use Sarah's superior grammatical constructions, even though they were not expressed in ASL, to upgrade my rating of chimpanzee language potential.

The lexical corpus with which we shall, therefore, deal may be summarized as follows:

John, born-deaf boy
    age 24 months: 83 ASL signs
    age 40 months: 134 ASL signs
Washoe, female chimpanzee
    after 22 months training: 34 ASL signs
    after ± 40 months: 92 ASL signs
Sarah, female chimpanzee (age or training times not available; age 6 when study began)
    58 signs in a special plastic token language
Sources: John [OLSON, 1972]; Washoe [GARDNER and GARDNER, 1969, 1971]; Sarah [PREMACK, 1971a, b, c]. Additional information from personal communications with OLSON, the GARDNERS, and PREMACK.

As I have indicated earlier, I am practically certain that if John's vocabulary had been determined by the same criterion as that used by the GARDNERS for Washoe, their performances at 40 months would have been much more nearly alike. It seems also likely that if both John and Washoe had been given equivalent training amounting to about one hour a day in the kinds of grammatical problems at which Sarah seems so expert, the outcomes again would have been more nearly alike. But it is unreasonable to expect perfection in two pioneer experiments or in the situation faced for the first time by the parents of a profoundly deaf child who must be trained to communicate in a strange new gestural system.

Semantic overlaps in the three cases may be summarized thus:

Terms unique to: John 82; Washoe 51; Sarah 47
Terms shared by: Washoe and John 4; Washoe and Sarah 4; John and Sarah 7
Terms shared by all three: 5

The small numbers of shared terms may seem strange. Reference to table I, however, shows that John and Washoe, particularly, had signs for many quite similar items, which simply reflects different household regimes. Thus, Washoe had signs for comb and blanket, whereas John had signs for pants and wallet; Washoe knew the sign for string, and John that for scissors, and so on. To bring this out more clearly for the entire corpus of lexical items, I have

Table I. Semantic clusters of lexical items used by Washoe, Sarah, and John. Washoe and John were trained in ASL (American Sign Language of the Deaf); Sarah was trained to communicate by means of plastic colored tokens

Clusters	Examples	Number of lexical items		
		Washoe	Sarah	John
Foods	banana, bread, water	5	17	19
Personal nouns and pronouns	I, you, mother	6		12
Adjectives, adverbs and prepositions	big, some, in	19	34	15
People	Greg, Sarah	8	7	1
Animals	dog, cat	5		12
Nature	tree, grass, wind	4		6
Parts of body	eye, ear	1		2
Clothing	coat, hat	7		6
Household objects	book, dish, bicycle	16	3	23
Colors	red, yellow	5	5	3
Verbs	look, eat, come	30	4	33

grouped the items roughly in semantic clusters based on my own subjective judgments. It must be remembered that the vocabularies presented for John and Washoe are not total inventories of all the signs they knew at the time, including those consistently recognized but not produced, or even those (in Washoe's case) which she knew and used, but did not use to the criterion adopted by the experimenters. They are in fact, vocabulary *samples.*

The names of persons listed in the examples column of the table refer to the research assistants and experimenters. John is reported by OLSON [personal communication] to have had no such name-signs by the end of OLSON's work with him; but it seems reasonable to suppose that now, at age five, he does use such signs. In ASL, personal names are commonly made by finger-spelling the initial letter (e. g., *W* for Washoe), followed by a sign referring to some personal attribute. Washoe's name, coined for her by the GARDNERS was *W* plus *ear*, because chimpanzees have very large ears. John's use of kinship terms must be regarded as functionally equivalent to the personal nouns and pronouns or the names employed by Washoe in ASL and in plastic token form by Sarah.

As with all instances of translation from one language system to another, absolute equivalence in meaning applications cannot be automatically assumed. Only fairly rigorous testing, such as was undertaken late in their pro-ect by the GARDNERS' [1971] use of sets of 35-mm color slides depicting members of a semantic class that differed in details (e. g., pictures of dogs of various breeds, sizes, and ages) can determine the extent to which a word is being used for all members of a class, or for a single, specific item. Such sets vary in their homogeneity; most raisins look alike, but dogs, cars, and trees may vary in many attributes; and it is safe to suppose that John's understanding of the term *city* would not coincide with that of an urban geographer or political scientist. *Tree*, for that matter, is not the name of any recognized botanical taxonomic category.

Inspection of the vocabularies reveals at once the heavy weighting of Sarah's experience toward acquisition of operator-terms which enable her to construct more elaborate sentences than John or Washoe. On the other hand, if placed in a household setting, Sarah would appear mentally backward compared to either, unable to symbolically differentiate *red book* from *red hat* or a cat from a picture of a cat from a picture of a pair of shoes.

In 1970, BRONOWSKI and BELLUGI expressed reservations about the ability of chimpanzees to acquire the rudiments of syntax; but the later reports about Sarah should have overcome those arguments. BROWN, also in 1970, wondered if Washoe could ever achieve the kinds of sentences which children make at comparable ages; but as we have indicated, it is hardly fair to compare an ASL-using subject with one using speech. The present data appear to narrow if not almost eliminate the gap in chimpanzee and human performance when the same sign-system is used by both. Such differences as remain between John and Washoe are comparable to normal differences in language acquisition in children with no determinable IQ deficit.

PREMACK [1971] has devoted an entire chapter in a recent book to the question of chimpanzee language competence, which goes a long way toward meeting the criticisms of BROWN and other specialists in human language acquisition. BROWN was impressed by the dearth of so-called *wh*-questions in the published material on Washoe; but aside from the evidence, e. g. from John, that such *wh*-questions (who, why, which, when, and where, etc.) are infrequent in ASL-using deaf children, there are other factors to be considered. Normal hearing children seem to use *wh*-questions as a kind of game when they become aware that most things have names and that their parents are willing to try to answer their endless questions. Since normal speaking adults can draw upon literally thousands of words, and are highly fluent users of their native languages, the game can proceed with little interruption, as long as the parent or baby-sitter is willing to play. Neither the GARDNERS nor the parents of John were fluent ASL 'native signers'. It would not take a very persistent child long to point to something for which he had no sign, an action that would force a conscientious adult to the sign-dictionary (where, by the way, the chances of finding an appropriate sign are much lower than they would be in a typical bilingual dictionary of spoken language). The *wh*-word game, according to recent students of ghetto or slum children and their language problems, seems to be played more rarely by poorly educated, overworked parents. Even in the affluent, there may come a point in the *wh*-word game, particularly when the interrogative is *why*, when the exasperated interlocutor answers, 'to make little boys ask silly questions', or falls back on 'just because, that's why'. End of game.

Negation was not easily elicited from Washoe, at least in ASL sign symbolic form; negative behavior was of course frequently exhibited. By the time a chimpanzee attains a body weight of 40 kg or so, he or she is so strong that unwillingness to comply with requests or commands can become a sheer test of muscular strength; whereas the human child, considerably weaker than any normal adult, must find symbolic negation a useful ploy. On one notable occasion, FOUTS [personal communication] relates that he had to concoct a little horror story to get Washoe to make the proper negation gesture. Operating at a level where use of negation only affected sentence construction, PREMACK seems to have encountered no peculiar barrier on Sarah's part to learning to say 'no' or 'is not'.

The principal area of last-ditch defense for the human-ape cognitive barrier, however, has to do with embedded or 'nested' sentence construction. This has become a favorite topic of modern transformational grammar specialists, and the protocols of chimpanzee sentences are in fact markedly poor in this respect. Complex embedding of clauses is hardly to be expected in conversations where few utterances are composed of more than three words. Sarah composed six- to eight-word sentences, but even these were two-part constructions linked by *and* or *if-then*. However, the evidence from normal speaking children shows that the ability to compose or to decode complexly embedded constructions is reached before five or eight years of age. Literary texts are the main source of our impression that human language characteristically employs such embedded constructions at high frequencies. In modern large language communities, there are also striking socio-economic class differences in the frequency of subordinate clause constructions. The linguists assume, perhaps rightly, that the ability to form such constructions is a human language universal, i.e., that examples could be found in all natural spoken languages. The data to prove such a dictum are insufficient, however, inasmuch as a great many of the more remote, isolated, and little-studied languages in the world are known today only by very incomplete wordlists.

I would guess that the ability to compose and to decompose sentences containing such embeddings is a relatively late human linguistic achievement unlikely to have been present in the earlier phases of language evolution. Word-games and conundrums like 'Brothers and sisters have I none, yet that man's father is my father's son' probably have existed for a long time; but I doubt if they were part of the language repertoire of *Homo erectus*. Ability to solve such conundrums probably exists, however, among the Australian aborigines, whose kinship terminology and marriage models would make the conundrum offered seem childishly simple. Until much more research is undertaken in this matter of embedding, I would certainly hesitate to claim (with apologies to those who prefer unmixed metaphors) that it represents the cognitive Rubicon, which no pongid can hope to cross.

Finally, there are philosophers of language who assert that a true language is one in which prevarication is possible. Since it is unlikely that honeybees can lie to their hivemates, so the argument goes, von Frisch is wrong; and bees lack 'language'. I find this a trivial point but am willing to argue that the ability to lie is not so much a matter of language use or abuse, but of demeanor. GOFFMAN [1971] has a good deal to say in a chapter, 'Normal appearances', on human dissembling. Prevarication or dissembling, if indeed absent among pongids[1], probably began to emerge in hominid behavior not only with hunting, where it is important for stalkers of game to pretend that they are not in active pursuit, but also in conjunction with the proliferation of incest taboos and similar sexual access-prohibitions. Lying is often expressed in language; but its essence seems to be an ability to conceal emotional cues, to suppress signs of one's intentions, or to fail to provide information useful to others. There is a large body of research on dishonesty in children and adults but little, if any, on nonhuman primates, as far as I know (see fn. 1).

PREMACK [1971b] ends one of his papers by remarking that man's uniqueness is still preserved since chimpanzees cannot teach language to man, but only man to chimpanzees. But that statement, seemingly true at first hearing, raises some questions. Feral children, brought up without any opportunity to acquire language from other human beings, might indeed be able to learn language from suitably trained ASL- or plastic-token-language-using chimpanzees, although the chimpanzees, it seems, would have had to get their language from human beings (and so into an infinite regression). If the state-

---

1   Cf. MENZEL, this volume. [Ed.]

ment means that chimpanzees, or whatever they might evolve into after several million years, could not be an imaginable starting point from which at last to derive language-using hominoids, I think the question must remain moot. An observer placed on this planet some time in the Pliocene might have found it absurd to consider whether the remote descendants of whatever early hominid ultimately gave rise to *Homo sapiens* would ever have achieved a propositional language.

I think that the experiments I have been comparing may teach us a great deal more than their authors intended about the threshold for language and that, if this line of investigation is pursued vigorously, we may be able to solve a great part of the puzzle of glottogenesis. Apart from that grand possibility, I think that chimpanzees, if well-equipped with a language like ASL, enriched by the kind of special grammatical training imparted by PREMACK to Sarah, may permit us to explore in a Piagetian fashion the higher reaches of cognition, an aspect of pongid behavior previously inaccessible because of the apparent language deficiencies of these animals in the Pre-Washonian era.

References

BRONOWSKI, J. and BELLUGI, U.: Language, name and concept. Science *168*: 669–673 (1970).

BROWN, R.W.: The development of Wh questions in child speech. J.verb.Learning verb. Behav. *7*: 279–290 (1968).

BROWN, R.W.: The first words of child and chimpanzee; in BROWN Psycholinguistics: selected papers, pp. 208–231 (The Free Press, New York 1970).

CICOUREL, A.V. and BOESE, R.J.: Cross-modal communication: the representational context of sociolinguistic information processing. Monogr. 25, Georgetown Univ. Monogr. Ser. Languages Linguistics (Washington 1972a).

CICOUREL, A.V. and BOESE, R.J.: Sign language acquisition and the teaching of deaf children. Amer. Ann. Deaf *117*: 27–33, 403–411 (1972b).

DENZIN, N.: Childhood as a conversation of gestures. Presentation. Ann. Meet. Amer. sociol. Ass., Denver (1971).

FOUTS, R.S.: The acquisition and testing of gestural signs in four young chimpanzees *Pan troglodytes*. Presentation. Ann. Meet. Animal Behav. Soc., Reno (1972).

GARDNER, B.T. and GARDNER, R.A.: Development of behavior in a young chimpanzee. 8th Summary of Washoe's Diary (University of Nevada, Department of Psychology, June 1970).

GARDNER, B.T. and GARDNER, R.A.: Two-way communication with an infant chimpanzee; in SCHRIER and STOLLNITZ Behavior of nonhuman primates, vol. 4, pp. 117–183 (Academic Press, New York 1971).

GARDNER, R.A. and GARDNER, B.T.: Teaching sign language to a chimpanzee. Science *165*: 664–672 (1969).

GOFFMAN, E.: Relations in public (Harper, Colophon Books, New York 1971).

Hewes, G. W.: An explicit formulation of the relationship between tool-using, tool-making and the emergence of language. Visible Language 8: 101–127 (1973).

Hewes, G.W.: Primate communication and the gestural origin of language. CA comments by R.J. Andrews, L. Carini, H. Choe, R. Gardner, A. Kortlandt, G. Krantz, G. McBride, F. Notterbohm, J. Pfeiffer, D.M. Rumbaugh, H.D. Steklis and M.J. Raleigh, R. Stopa, A. Suzuki, S.L. Washburn, and R.W. Wescott: Current Anthrop. 14: 5–24 (1973).

Hill, J.H.: On the evolutionary foundations of language. Amer. Anthropologist 74: 308–317 (1972).

Lieberman, P.; Crelin, E.S., and Klatt, D.H.: Phonetic ability and related anatomy of the newborn and adult human, Neandertal man, and the chimpanzee. Amer. Anthropologist 74: 287–307 (1972).

Mattingly, G.: Speech cues and sign stimuli: an ethological view of speech perception and the origin of language. Amer. Scientist 60: 327–337 (1972).

Menzel, E.W.: Spontaneous invention of ladders in a group of young chimpanzees. Folia primat. 17: 87–106 (1972).

Olson, J.R.: A case for the use of sign language to stimulate language development during the critical period for learning in a congenitally deaf child. Amer. Ann. Deaf 117: 389–396 (1972).

Pepys, S.: in Latham and Mathews Diary, vol. 3 (1962) (University of California Press, Berkely 1970).

Peters, C.R.: Evolution of the capacity for language. Man, new Ser. 7: 33–49 (1972).

Premack, D.: Language in chimpanzee? Science 172: 808–822 (1971a).

Premack, D.: Some general characteristics of a method for teaching language to organisms that do not ordinarily acquire it; in Jarrard Cognitive processes of nonhuman primates, pp. 47–82 (Academic Press, New York 1971b).

Premack, D.: On the assessment of language competence in the chimpanzee; in Schrier and Stollnitz Behavior of nonhuman primates, vol. 4, pp. 185–228 (Academic Press, New York 1971c).

Riekehof, L.: Talk to the deaf (Gospel Publishing House, Springfield 1969).

Skinner, B.F.: Verbal behavior (Appleton-Century-Croft, New York 1957).

Stokoe, W.C., jr.: Sign language structure: an outline of the visual communication systems of the American deaf; in Studies in linguistics. Occasional Paper No. 8, p. 78 (State University of Buffalo, Buffalo 1970).

Trân Dûc Thâo: Le mouvement de l'indication comme forme originaire de la conscience. La naissance du language etc. Pensée 12: 3–24; 147,148,149: 903–1106 (1966–1970).

Wiener, M.; Devoe, S.; Rubinow, S., and Geller, J.: Nonverbal behavior and nonverbal communication. Psychol. Rev. 79: 185–214 (1972).

Yerkes, R.M.: Almost human (The Century Company, New York 1927).

Author's address: Dr. Gordon Hewes, Department of Anthropology, University of Colorado, Boulder, CO 80302 (USA)

Symp. IVth Int. Congr. Primat., vol. 1: Precultural Primate Behavior,
pp. 144–184 (Karger, Basel 1973)

Cultural Elements in a Chimpanzee Community[1]

JANE VAN LAWICK-GOODALL

Gombe Stream Research Center, Tanzania, and Stanford University, California

The anthropologist, in his investigations into the social behaviour and customs of some human group or population, typically has an all-embracing concept of culture. Whilst he may acknowledge that smiling is an innate movement pattern, occuring as it does in all human peoples known today and even in children born blind [EIBL-EIBESFELDT, 1970], he is perfectly correct in his assumption that custom may determine the precise manner of smiling or the contexts in which it is most usually given. He is, in other words, investigating 'that complex whole which includes knowledge, belief, art, morals, law, custom and any other capabilities and habits acquired by man as a member of society' [TYLOR, 1871]. There are customs and taboos concerning almost every aspect of behaviour from religious beliefs to menstruation, from the way in which music is made to the manner of weeping and the contexts in which it is permissible to weep. Therefore, in essence, all behaviours observable to the anthropologist studying human beings can be described as a part of their culture.

The ethologist, on the other hand, if he is attempting to investigate 'culture' in a nonhuman animal species, is primarily interested in the careful analysis of behaviour patterns and their ontogeny. Only a few elements of behaviour may be considered, after careful scrutiny, to have been influenced by culture. And whether or not they can be thus labelled depends on the manner in which they were acquired by the individual, i.e., the kind of learning process involved. The anthropologist studying human beings is mainly interested in the material which comprises a culture; whilst the ethologist is principally

1   This work has been made possible by a generous grant from the Grant Foundation of New York.

concerned with the precise manner in which particular patterns are transmitted from one individual to another in a group.

HANS KUMMER [1971] gives us a clear and concise account of the ethological concept of cultural behaviour in the nonhuman primate. Species adaptation, he points out, occurs in two major ways: (1) through phylogenetic adaptation, the slow, gradual evolution of the genotype of the species; and (2) through ontogenetic adaptation[2], the adaptation of the individual to the particular environment in which it finds itself. Ontogenetic adaptation can also be sub-divided into two main categories: (1) ecological modification, which results from adaptation to such factors as climate, terrain, types of predator, and so on; and (2) social modification, which results from the influence of other individuals in the social group. 'If such social modification spreads and perpetuates a particular behavioural variant over many generations, then we have "culture" in the broad sense in which a student of animals can use the term.' KUMMER goes on to define cultures as 'behavioural variants induced by social modification, creating individuals who will in turn modify the behaviour of others in the same way' [KUMMER, 1971, p. 13].

The fundamental problem in applying this concept to a primate society lies in the difficulty of determining precisely which behaviours are 'variants'. KUMMER's solution lies in a knowledge of the behaviour of the two groups of the same species, with the same gene pool, living in similar environments. If their behaviour *differs*, then the deviance must be cultural [KUMMER, 1971]. But it does not necessarily follow that because a certain behaviour is the same in two groups that it is *not* culturally acquired; if we could look further back into the history of the species, we might find that a cultural modification had, in fact, taken place.

It seems then that before we can be certain that a behaviour is culturally rather than genetically determined we may need to study many groups. In addition, the data must be collected in essentially the same way; and the behaviours in question must be examined rigorously for possible differences in expression or context that might point to otherwise unsuspected deviations between groups. This approach, for instance, might reveal that whereas all chimpanzee infants start to ride on their mothers' backs at a similar age throughout the range of the species, the mothers in different areas might show slightly different ways of initiating the behaviour. Such differences in maternal technique could probably be called cultural, but there would still be the possibility that at least in some populations genetic factors might be at work.

2   This is my terminology; KUMMER uses 'adaptive modification'.

Unfortunately, we do not, at the present time, have the kind of data on chimpanzees that would make this approach possible. Therefore, for the purpose of the present discussion, whilst accepting Kummer's definition in principle and making as much use of the comparative approach as possible, I shall primarily be concerned with an examination, in one group of wild chimpanzees, of those aspects of behaviour that seem *most likely* to be culturally influenced. Much of the material will, therefore, be speculative, but may help to stimulate further research in this field.

The Chimpanzee Society at Gombe National Park

Before I can discuss culturally influenced elements of behaviour it is necessary to give an outline of the social structure of the chimpanzee community at Gombe Stream [for fuller accounts, see van Lawick-Goodall, 1968, 1971, 1973]. This community, studied since 1960, lives in a narrow stretch of rugged, mountainous country on the eastern shores of Lake Tanganyika in Tanzania. The Park is approximately 30 square miles and supports four, or possibly five, more or less distinct communities of between 20 and 40 chimpanzees. (The community is the equivalent of Nishida's unit-group [Nishida, 1968] and probably the same as Sugiyama's 'regional population' [Sugiyama, 1969] and the 'large-sized group' of Itani and Suzuki [1967].)

Within the community, individuals move about, for the most part, in small temporary associations, membership of which is constantly changing as individuals or groups of individuals separate to move about alone or to join other associations. These groups may be all male, female and young, adult male and adult female, or combinations of all or any of these age-sex classes. Occasionally, almost all the individuals of a community will join together to feed on some specially favoured food. Males and sometimes females may move about for hours or even days alone. Some individuals in the community meet up only when chance, such as a local abundance of food or a female in oestrus, throws them together; others meet more often; some show strong bonds of mutual attraction and travel, groom, feed, and rest together very frequently. A mother and her dependent offspring remain constantly together over a period of years. We do not yet have much information relating to encounters between individuals of different communities, but we do know that some chimpanzees may penetrate the home-range of a neighbouring community and peacefully associate with at least some of its members.

Occasionally, a female may leave her natal group and transfer to a new community.

A complete description of the social behaviour and way of life of this chimpanzee community would, in some ways, be analogous to the anthropologist's description of the 'culture' of a group of humans (although I am not trying to suggest that all the chimpanzee behaviours are culturally acquired). Suffice it to say here that the community has a home-range of approximately 20 square miles within which the chimpanzees are nomadic to a large extent. They build sleeping nests in the trees at night, close to the spot where dusk finds them. They are omnivorous, feeding on plant material, insects, birds' eggs, and occasional fledglings. From time to time they hunt medium-sized mammals (such as monkeys or young bushbuck) and on such occasions may show quite sophisticated cooperation and food sharing. The chimpanzees use a variety of objects as tools in a wide variety of feeding and other contexts; occasionally, they modify such objects and thus show the beginnings of tool-*making* behaviour [VAN LAWICK-GOODALL, 1970]. Chimpanzees communicate with each other by means of an extremely sophisticated and complex repertoire of vocal, postural, and gestural cues, the detailed study of which is only just beginning at Gombe.

In the wild, the period of childhood is exceptionally long. The infant does not start to walk until he is six months old; he continues to nurse, travel from place to place on his mother's back, and sleep with her in her nest at night until he is five or six years old. The juvenile male may begin to travel independently of his mother for a few hours at a time when he is about seven, but he does not leave her for more than a few days until he is about ten. The female may associate closely with her mother for longer. Indeed, we know now that bonds between a mother and her offspring are very persistent and may continue throughout the life of the mother. Puberty occurs at approximately eight or nine years of age and is followed by a period of adolescence. Social maturity is attained by the male at about fifteen years of age. The female may have her first infant when she is about thirteen; she has only one infant every four to six years.

Since HENRY NISSEN made his pioneering study [NISSEN, 1931], several different chimpanzee groups have been studied. Those studies to which I shall mainly refer in this paper are summarized in table I. A careful examination of this literature suggests that a similar kind of social structure and similar kinds of behaviours to those observed at Gombe are typical for the species throughout its range. For the most part, cultural differences between various study groups must be determined in the future.

Experience and Learning

The manner in which a given behavioural trait has been acquired by an individual may, as we have seen, be helpful in determining whether it may be regarded as cultural. It is, therefore, necessary to discuss briefly the role of learning and experience in the ontogeny of chimpanzee behaviour. As the mammalian brain becomes increasingly complex, culminating in the brains of the higher primates, learning and experience certainly play an ever more vital and significant role in the life of the individual. This statement is certainly true for the chimpanzee; at the same time we must not forget that many of his movements and postures have been genetically programmed during the course of evolution so that he is often predisposed to behave in a certain manner. Thus, MENZEL [1964], found that entire sequences and patterns of behaviour, as well as many responses, were latent in chimpanzee infants that had been raised for the first two years of their lives in social isolation and in restricted environments. He also comments that the 'sheer non-specificity, interchangeability, and plasticity of releasing stimuli' was impressive. In other words, although species-specific action patterns may appear, it is necessary for the chimpanzee to learn a great deal about the contexts in which they should be used (see ROGERS, this volume).

1. Learning by Experience

In the wild, the infant chimpanzee learns much during his day-to-day experiences with his physical and social environment, an environment which has gradually produced over thousands of years the genotype for the chimpanzee. Early grasping of branches and locomotor movements of the limbs will result in typical climbing behaviour; the coordination and skill of the youngster in climbing will increase as a result of maturation, practice, and growing familiarity with the physical properties of different kinds of tree trunks, branches, and vines.

Just as basic locomotor patterns may be inherited through the genes, so too are some of the components of the gestural, postural, and vocal communication system of the chimpanzee. As the infant grows older, he will begin to direct movements or sounds of this sort towards his mother and other individuals in the group. As a result of social reward for some of these spontaneous actions, they are likely to be repeated so that his repertoire of species-specific gestures and calls gradually emerges, patterned and organized into sequences as a result of his experiences with others.

Initially, the behaviour of the infant is molded by clear-cut signals of his mother. When she is ready to move off, she presses him firmly to her breast or picks him up and places him on her back. If he totters towards an adult male who shows signs of aggression, his mother will hurry after him and carry him to safety. Gradually, however, the infant learns to respond to the gestural, postural, or auditory equivalents of his mother's early signals. When she wants to move, she need only tap the nine-month-old infant lightly and he will climb aboard. And when he is older still, a slight gesturing, a soft call, or simply a glance in his direction may be sufficient cues to convey the same information. When he is about a year old, he himself is able to recognize signs of aggression in an adult and will hurry out of danger of his own accord. A small infant who closely approaches a feeding male is normally tolerated; at most he will be gently held off. The four-year-old, however, is likely to be pushed away more roughly; later still he may be hit. Eventually he learns either to avoid some feeding males, or to respond rapidly to very slight threat gestures.

Some of these processes of socialization may function in a similar way to the early disciplining of a human child. The infant chimpanzee learns, partly through the guidance of his mother, partly through his own social experiments, which behaviours are acceptable in his society and which ones may lead to trouble for himself. He gradually becomes familiar with the individual idiosyncrasies of behaviour of other individuals; e.g., how a submissive gesture to male A will normally lead to a reassuring contact, whilst the same gesture directed in a similar context to male B may result in eliciting a mild threat. Gradually he learns, too, how the status of other individuals may change with relation to their companions at the time: a juvenile whom he can threaten with impunity when she is on her own may attack him vigorously for the same behaviour if her mother or elder sibling is nearby to back her up.

2. Social Contagion (or Facilitation) and Local Enhancement

Sometimes a certain behaviour may be contagious in a group of social animals. In chimpanzees, for example, there are certain calls which, if they are uttered by one individual, are likely to induce others present to do the same. From an early age, the infant is likely to join in when adults nearby start to 'pant-hoot'. As a result of repeatedly joining in such choruses, he gradually learns the appropriate contexts for the call. In a similar way when he is older, if he joins in a few times when adults are mobbing a predator or himself flees as a result of seeing his companions in flight, he will learn something of the dangers of the environment and the manner in which his society typically responds to them[3].

Learning can also occur when the behaviour of a companion serves to direct the attention of the individual to certain aspects of the environment. For instance, an infant chimpanzee, during his years of dependency, accompanies his mother time and again along certain routes which she prefers. Subsequently, when he begins to travel independently, he is likely to choose many of these tracks himself.

3. Observational Learning

Chimpanzees, along with some other animals, are able to learn as a direct outcome of observing the manner in which a particular behaviour is performed by another individual. A rhesus macaque monkey, for example, can readily learn the correct response from watching another monkey performing in a discrimination experiment. Moreover, the observing animal is able to benefit from incorrect as well as correct responses of the performer; he learns as a result of the consequences of the other's actions [Darby and Riopelle, 1959]. The success of the chimpanzee in observational learning tasks has been well documented for captive individuals, both in experimental situations [e. g. Yerkes, 1943], and in home-raised chimpanzees [e. g. Hayes and Hayes, 1952; Gardner and Gardner, 1969].

Learning through imitation of the behaviour of other chimpanzee models is undoubtedly of great significance in the development of the infant in the natural habitat and may, on occasion, be important also for the adult. Often the types of learning mentioned here are closely interrelated in the natural course of things. Thus initially, the attention of the infant chimpanzee may be directed towards some aspect of the environment, such as a certain type of food, because he sees a companion feeding. He may become highly motivated to eat because the sight of another feeding can be contagious. Then, if the food in question is difficult to obtain (such as insects in an underground nest or fruits with rinds difficult to break), he may carefully watch the way in which the model handles the problem and imitate the behaviour.

---

3   Social contagion, or social facilitation, does not necessarily imply learning; yawning, for example, is very contagious.

4. Insight

One additional point should be made. When faced with some novel (to himself) problem, whether it concerns a method of obtaining a fruit hanging beyond his reach or of obtaining a share of the food of a social superior, the chimpanzee may attempt a solution using the motor patterns and intelligence available to him. If one of his attempts succeeds, he will not necessarily perceive the connection between that particular action and the desired result. When that problem crops up again, he may go through a number of the same unsuccessful procedures before once more hitting on the right solution. After two or three such incidents, however, it is likely that the lesson will be learned; on the next occasion he will immediately perform the appropriate action. In addition to this trial and error learning, chimpanzees may, when faced with a difficult problem and after an initial failure to solve it, suddenly show purposeful action and instantly solve the task. In other words, they have the ability to perceive certain relations between things of the external world. This kind of 'insight' or 'ideation' has been well described by many people who have worked with chimpanzees, especially by KÖHLER [1925] and YERKES and YERKES [1929]. It may be of great importance in the development of new cultural elements in a species in which observational learning also plays a vital role. As a result, one individual's insight may afford the opportunity to many individuals to learn a behaviour that the majority of them might never discover on their own. It is, after all, due to the genius of certain gifted individuals of our own species that man has been able to make such tremendous technological strides in recent decades.

The foregoing discussion gives some idea of the complex role played by experience and learning during the development of a chimpanzee in nature and suggests some of the ways in which behaviour may be modified by conspecifics. The task of trying to unravel elements of behaviour that have *not* been, to some extent, culturally acquired, particularly in the sphere of social relationships, is a difficult one. Indeed, until we know more about details of chimpanzee behaviour in other areas, it is wiser not to attempt the job. In this paper I shall, therefore, concentrate on those elements of behaviour that appear to be the most obviously influenced by observational learning and imitation and also discuss some of the pathways along which behaviour might be transmitted from one individual to another through successive generations. It is convenient to make an arbitrary division and discuss firstly behavioural elements in the sphere of ecological adaptation, such as feeding and tool-using, and secondly, elements in the sphere of social relationships, e.g. the postures and gestures of the communication system.

Cultural Elements in Ecological Adaptation

1. Nest-Making

Nest-making behaviour probably occurs throughout the range of the chimpanzee, but the extent to which observational learning is necessary for the development of the pattern is not yet known. Preliminary experiments, however, show that laboratory-born chimpanzees, separated from their mothers after a few days, never make nests in situations where wild-born individuals will do so [BERNSTEIN, 1962]. Certainly, infant chimpanzees in the wild watch when their mothers construct nests. Twice at Gombe small infants were observed 'helping' their mothers to construct the communal bed by reaching out and bending twigs over. From the age of about eight months, infants gain experience by themselves in constructing small nests, often as a form of play activity during the day.

2. Feeding

There are two major ways in which feeding behaviour might be influenced by culture: (a) the kind of foods selected; and (b) the manner in which a given item is obtained or prepared for feeding. Extensive collections of food-plants have been made and identified in both the major Tanzanian study sites, the Kasakati Basin area and the Gombe National Park (see table I). A rough appraisal of the data suggests that foods selected in the one area are also selected in the other [R. WRANGHAM, personal communication]. Further work needs to be done before we know whether some of these items are more highly favoured in one area than in the other. Nor can we yet attempt to elucidate (b) from the existing literature.

We do know that in some areas, where there are local plantations of grapefruit or papaya, chimpanzees will feed on them [ALBRECHT and DUNNET, 1971; KORTLANDT, 1962]. At Gombe, where there are no such plantations, we have offered both these types of fruits and they have consistently been refused. Other unfamiliar foods have also been rejected. This response is not surprising since adult chimpanzees are known to be very conservative in their feeding

Table I. Major field studies of the chimpanzee throughout its range

Locality	Year	Duration of study	Investigators
Guinea: Kanka Sili area	1930	2½ months	H. NISSEN
	1968/69	several months	A. KORTLANDT
			H. ALBRECHT
			S. DUNNETT
Eastern Congo: Beni	1960/61	several short field studies	A. KORTLANDT
Western Uganda: Budongo Forest	1962	9 months	V. REYNOLDS
			F. REYNOLDS
	1966	9 months	Y. SUGIYAMA
			A. SUZUKI
Western Tanzania Kasakati Basin	1964 to present	8+ years	J. ITANI
			K. IZAWA
			T. NISHIDA, M. KAWABE and members of the Japan Monkey Centre
Gombe National Park	1960 to present	12+ years	J. VAN LAWICK-GOODALL and members of the Gombe Stream Research Centre

References: NISSEN, 1931; ALBRECHT and DUNNETT, 1971; KORTLANDT, 1962, 1963, 1965, 1968; REYNOLDS and REYNOLDS, 1965; SUGIYAMA, 1969; ITANI and SUZUKI, 1967; IZAWA and ITANI, 1966; SUZUKI, 1966, 1971; NISHIDA, 1968, 1970; NISHIDA and KAWANAKA, 1972; KAWABE, 1966; VAN LAWICK-GOODALL, 1968, 1970, 1973.

habits. This kind of adult conservatism is characteristic of a variety of primate species; KUMMER [1971] has pointed out its adaptive advantage: it may be dangerous to experiment with new foods; therefore, if some individuals are resistant to changes in diet, the group would survive even in the event that some members were poisoned. In Japanese monkeys *(Macaca fuscata)*, where adults are also resistant to new foods, it has been possible to document the influence of young monkeys in the gradual transmission of a new food preference through at least part of the group [e.g. ITANI, 1965; KAWAI, 1965].

We have not, as yet, observed the diffusion of a new food habit through the Gombe chimpanzee community; but a few observations suggest that in chimpanzees also it is young animals that would be the most likely to initiate such a habit. When infants are exploring the environment, they frequently place in their mouths a variety of plant material that adults have not been seen to feed on. The infants have not been seen to eat such items, but they do have the opportunity to taste them. A nine-month-old infant repeatedly sniffed and licked a mango skin lying on the ground, and a four-year-old male was once seen to eat a whole mango. (That the mango is not part of the diet of the Gombe chimpanzees, despite the fact that mango trees have been growing in the area for at least 60 years, is a dramatic illustration of chimpanzee conservatism in feeding habits.) From time to time we have offered many of the chimpanzees various types of new foods; but only twice have such items been accepted: a seven-month-old infant ate a tiny piece of biscuit, and a juvenile briefly chewed on some sugar cane. On several other occasions infants touched, smelled, and licked at foods offered them; and adolescents sometimes picked up such items and smelled them carefully. Adults, however, usually ignored novel food items; occasionally, they bent to smell them briefly.

Thus it appears that young chimpanzees are typically more exploratory and adventurous in trying new food items. Often, however, their experiments in feeding may be quickly discouraged by their mothers or other individuals. When a year-old infant began mouthing the fruit of a plant not listed among the food items of the Gombe chimpanzees, his mother repeatedly tried to flick it from his mouth with her finger [R. Wrangham, personal communication]. A three-year-old picked up a piece of papaya and began to lick it; her mother at once snatched it away and sniffed it carefully before discarding it. And, on the occasion when the small infant ate a piece of biscuit, his elder sister hurried up quickly, sniffed the remainder of the food, and then vigorously flicked it away from him. Intervention of this sort undoubtedly serves to maintain the traditional food preferences of the community.

Moreover, despite this occasional flexibility in the feeding behaviour of youngsters, they are for the most part far more likely to feed on the food which their companions are eating than to conduct their own experiments. An infant chimpanzee, from the age of about five months or even younger, often watches closely when his mother is feeding. He may also put his face very close to hers as though sniffing the food she is eating. Once the infant begins to eat minute portions of solid foods in addition to his milk diet, he normally chews at the same food that his mother is feeding on while she is eating. He is also likely to sample foods eaten by his siblings, play-mates, or other individuals who happen to be in the group.

When the infant is about two years old, his repertoire of known foods (those he has eaten as a result of watching the feeding behaviour of others) has increased. He is now more likely to feed independently, particularly if his mother is consuming a type of food which he finds difficult or impossible to eat himself, such as a seed with a tough pod or a fruit with a hard rind. Equally often, he becomes interested in the behaviour shown by his mother when she is eating such foods; and he may watch her closely, beg from her, and sometimes make ineffectual attempts to imitate her method of coping with the situation.

One such food is the hard-shelled *strychnos* fruit, about the size of a tennis ball, which the adult opens by banging against the trunk of a tree or rock. Infants under four years of age are seldom able to crack the rind and obtain morsels only by begging from others. After watching her sibling opening such a fruit, a two-year-old then hit her own wrist against the tree trunk. Three-year-olds frequently attempt to bang the fruits themselves against tree trunks, though they usually drop them after a few attempts. A six-year-old male finally managed to open a strychnos fruit after making no less than eighty bangs, as compared with the half dozen or so of the adult [B. WRANG-HAM, personal communication].

Chimpanzee infants probably learn to eat biting or stinging insects by watching adults. Three kinds of ants with painful bites are eaten, and bees' nests are raided. Infants under about five years normally avoid getting too close when such items are being eaten, but they usually watch from a safe distance. Weaver ants *(Oecophylla longinoda)* construct a nest by joining a clump of leaves together with sticky silk; they are eaten by chimpanzees during about two months of the year. An adult chimpanzee will pick an entire nest and then make a rapid downward movement with one hand or foot which serves to sweep off any ants crawling on the surface and probably also partially to crush those within. He then breaks open the nest and feeds on the insects. One infant, about three years of age, was not observed to attempt to feed on weaver ants during the season. The following year I saw her, after watching an older playmate feeding on these insects, hurry to a hanging nest, pick it very quickly, make a few frantic-looking sweeping movements, race from the spot, make a few more sweeping movements, and then drop the entire nest without eating any of the ants.

3. Tool-Using

Other items of the diet of the adult which are difficult for the infant to acquire, including other biting insects, are those obtained by the use of an object as a tool. Termites *(Macrotermes bellicosus)*, which form a major part of the diet for three to four months each year, are 'fished' out of their underground nests with the aid of grass stems or slender twigs. This behaviour has been described in detail elsewhere [van Lawick-Goodall, 1968, 1970]. The chimpanzee opens up a nest passage by scratching at the covering layer of soil with a finger or thumbnail, carefully pushes down a fairly long tool, pauses for a moment, withdraws it, and picks off with his mouth any termites that adhere. If he picks a leafy twig for use, he will strip off the leaves; if he picks a blade of grass that is too wide, he will trim the edges. When his tool breaks or becomes bent or too short, he selects a new one.

The small infant, before he ever attempts to use a tool in the correct context, frequently watches his mother as she works. Occasionally he may scratch at the surface of the heap with his index finger, or he may poke or prod at a termite crawling past. By the time he is seven or eight months old he may occasionally eat a termite, usually one that is on his mother's hand or wrist. Moreover, although we do not yet have quantitative data to support this statement, it certainly seems that during the termiting season small infants begin to use grass stems more frequently as play objects than before. Certainly they very often break off grasses, or pick them from the ground, and play with them whilst their mothers are actually working at a termite heap. They may also pull at or bite the end of grasses that the mothers are using as tools.

During the second month of his first termite season, when an infant was eight months old, he incorporated 'mopping' into his gestural repertoire. Mopping is a movement that is rarely if ever seen out of the termiting context and occurs when a number of the insects are crawling about on the surface of the nest. The adult gently places the back of hand and wrist over the termites and gently rotates the hand laterally. This motion causes the termites to become entangled in the hair, or they may bite onto it. The chimpanzee then picks them off with his lips. At first this infant just made banging-down movements onto objects with the back of his wrist, but after a week he also showed a slight outward rotation of the hand. He did not show the behaviour in context, but mopped almost anything, branches of trees, the ground, rocks, his mother's leg. Occasionally he also mopped at the surface of a termite nest, but was never seen to direct the gesture onto a termite.

An infant between one and two years of age often picks a small twig whilst his mother is working at a nest and, apparently playfully, strips off the leaves, thus imitating the manner in which an adult prepares a tool for use. Two infants of one-and-a-half years old, after watching their mothers working, picked up short thick pieces of stem and jabbed them at the surface of the nest. In neither case was there a hole there. One of these infants held her 'tool' in the manner of an adult, between thumb and forefinger. The other used the 'power grip' [NAPIER, 1961], holding her piece of stick rather as a human infant may initially hold a pencil or spoon. From two years of age the infant begins to show more adult patterns, although he will not normally develop the delicate manipulative techniques of the adult, nor consistently select and use appropriate materials, nor open up good passages for himself until he is five or even six years old. Small infants almost always use tools that are too short or too thick or too flexible [VAN LAWICK-GOODALL, 1970].

Since it occurs very frequently, termite fishing behaviour offers an excellent opportunity for studying the ontogeny of the tool-using behaviour in infants. Without doubt, observational learning also plays a significant role in the acquisition of the other tool-using techniques of the Gombe chimpanzees. Ants of two species are eaten usually with the aid of quite large sticks. When raiding an underground nest of *Anomma nigricans* (the 'safari ant'), the chimpanzee plunges his stick down into the nest (usually from an overhead branch), waits for a moment, withdraws the stick, and sweeps his other hand along its length. He thus collects a handful of ants, which he eats. Once, when a five-year-old saw a few of these ants moving over the surface of the ground, she broke off a stick, pushed it down into the sandy soil, stepped back, and then pulled out her tool. Since there was no nest at that spot, her behaviour was unrewarded. Another species of ant, *Crematogaster* sp., constructs a hard-walled nest on the branches of trees. Chimpanzees may push sticks into these nests and pick off the ants. One mother carefully examined an intact nest, selected a stick, and using it as a lever, tried unsuccessfully to push it between the nest and the branch. After watching, her five-year-old daughter also found a stick and tried to use it in the same way.

The Gombe Stream chimpanzees use leaves as a 'sponge' to sop up rainwater from a hollow in a tree trunk or branch that they cannot reach with their lips. Before use the tool is modified: the chimpanzee briefly chews the leaves, thus crumpling them and increasing their absorbency. All of the adults who have been observed drinking from such water bowls have used leaves in this way. Some infants, however, have merely dipped their hands into the bowl and licked the water from their fingers. F. PLOOY [personal communication] watched as one mother began to use a sponge. Her two-year-old daughter moved up to sit closely behind her, but was unable to get a good view of the activity from this position. Since the mother was sitting on the only branch projecting from the main trunk at that place, the infant made a two-minute journey (with pauses) involving locomotor manoeuvres difficult for her, until she reached a place from which she could watch her mother's behaviour. She looked intently for half a minute and moved away until her mother left the water bowl. The infant then returned and for six minutes repeatedly dipped her hand in the water and licked her fingers. Another infant of similar age watched her mother drinking with leaves and then used the sponge which had been left in the bowl. Infants between three and four years of age have success-fully demonstrated the adult technique.

Leaves are also used to wipe dirt or blood from the body. On two separate occasions a two-year-old male, after watching his mother pick handfuls of leaves to wipe diarrhoea residue from her bottom, picked leaves to wipe his own clean bottom. In neither case had he himself defecated.

Quite frequently a chimpanzee will use a grass stem or a stick as an 'olfactory aide' or investigation probe, in order to touch something he cannot reach or fears. A thin twig may be pushed into a hole in a piece of rotten wood. The chimpanzee withdraws it, sniffs the end, and then either moves away or else breaks open the wood. Usually this latter action reveals an insect larva, which is promptly eaten. One juvenile female, after sitting with a group of adults staring at a dead python, pushed a long dead palm frond, hand over hand, until the tip touched the python's head. She withdrew her implement, sniffed the end, and then repeated the process twice.

Table II lists the extent to which the tool-using behaviours observed at Gombe have been recorded for chimpanzees in other areas. I emphasize that the lack of observation of a tool-using performance in a given locality cannot be taken to imply that it does not occur. It was several years before the listed repertoire was recorded at Gombe. On the other hand, it would be surprising if, after 12 years of intensive study, all of the *frequently shown* tool-using patterns of the Gombe chimpanzees had not been recorded. It is, therefore, reasonably safe to say that these chimpanzees do not normally use leafy branches as fly-whisks [Sugiyama, 1969] or rocks to hammer open hard-shelled fruit [Beatty, 1951; Savage and Wyman, 1843/44]. Certainly Gombe chimpanzees have been observed in situations when fly-whisks would have been most adaptive, such as when a mother for a couple of days carries around the body of a dead infant, which attracts such great swarms of flies that she is constantly waving them off with her hand.

The fruit of the oil-nut palm is a staple food of the chimpanzees at Gombe, but we have never seen an individual feeding on the kernel; to do so, a rock would certainly have to be used to break open the excessively hard shell. It is of interest that an infant at Gombe was once observed to hit repeatedly with a

Table II. Extent to which tool-using behaviours observed at Gombe have been seen in other areas and vice versa

Object	Use	Gombe	Kasakati	Uganda	Congo	Guinea	Other
Grass stem	termite feeding	+	+ [1]				Gambia¹
Grass stem or stick	honey feeding	+	+ [2]				Cameroons [11]
	investigation	+				+ [9,10]	Gambia¹
Stick	ant feeding	+					
	lever	+					Gambia¹
	enhance display	+	+ [3]	+ [4]	+ [6]	+ [10]	
	weapon	+			+ [6–9]	+ [9,10]	
Leaves	wiping body	+					
	drinking	+		lick water from hands [4]			Gambia¹
	fly whisk (on twig)			+ [5]			
Rock or stone	enhance display	+	+ [3]			+	
	missile	+				+ [10]	
	hammer	1 infant					Liberia [12]
	open fruit	hit object					Central Africa [13]

References: [1] SUZUKI, 1966; [2] IZAWA and ITANI, 1966; [3] NISHIDA, 1970; [4] REYNOLDS and REYNOLDS, 1965; [5] SUGIYAMA, 1969; [6] KORTLANDT, 1962; [7] KORTLANDT, 1968; [8] KORTLANDT, 1963; [9] KORTLANDT, 1965; [10] ALBRECHT and DUNNETT, 1971; [11] MERFIELD and MILLER, 1956; [12] BEATTY, 1951; [13] SAVAGE and WYMAN, 1943/44.

1 Crude termite-fishing behaviour and sophisticated use of a leaf sponge, as well as the use of sticks for investigation and levers, have been reported in a captive group of chimpanzees that is allowed to range freely by day in a forest in the Gambia. These chimpanzees were obtained as orphans, probably from Portuguese Guinea, and had been with their mothers for two to three years before capture. Thus they probably show tool-using behaviours typical of their original society [BREWER and BREWER, personal communication].

stone at some unidentified object on the ground (probably an insect). This shows that the 'hammering' technique is available. It was probably from such chance performances, which also happened to yield a worthwhile reward, that some of the typical tool-using techniques have developed; the adaptive 'invention' of one individual would probably spread through the community as a whole through observation and imitation.

One factor which is possibly crucial in the evolution of tool-using performances is that a chimpanzee infant, once he has mastered a complex manipulative or motor pattern, practices it quite frequently. Moreover, as we have already seen in the example of the 'mopping' behaviour of one infant, the releasing stimuli can be very diverse. There are many other examples illustrating this point. A two-year-old, during the termiting season, three times pushed a piece of grass carefully through the hairs of his own leg, and once into the groin of his sibling. He used a termite-fishing technique and, after withdrawing his 'tool', put the end to his lips. Another small infant used a tiny leaf sponge to dab at sticky fruit pulp which he had accidentally smeared onto the branch of a tree. 'Inspection' is a typical response to a female who has presented her rump in a submissive context; it is ordinarily performed with a finger, which the chimpanzee then sniffs. Two infants, however, each used a small twig to inspect a female. A five-year-old used a twig to investigate his newborn sibling after his hand had been repeatedly pushed away by his mother.

Thus, as is the case with so many behaviours in infancy, tool-using patterns have not yet become as context-bound as in the adult. As a result, one tool-using pattern used in a different context may lead to a novel adaptive performance for the individual concerned. One three-year-old, for example, used a blade of grass with the termite-fishing technique when trying to drink from a water bowl. Each time after he withdrew his inappropriate tool and sucked the end, the grass became more crumpled until eventually he had fashioned a tiny sponge. He continued to use this for a short while before losing interest. This example not only gives us some insight into the way in which the use of a sponge may have originated in chimpanzee history, but also shows that individuals may, in fact, 're-invent' some of the tool-using techniques of their society.

4. 'Leaf-Grooming'

There is another behaviour pattern shown by the Gombe chimpanzees that merits inclusion in this section. Sometimes, particularly when a chimpanzee is involved in a social grooming session, he will reach out and seize a leaf or several leaves. Peering closely at the leaves and often lip-smacking, he holds them with both hands and makes clear-cut grooming movements with one or both thumbs. Sometimes he will remove minute specks with his lips. After a few moments he will drop the leaves and resume other activities. Nearly always this 'leaf-grooming' activity arouses great interest in nearby chimpanzees, who may cluster closely around the individual concerned, peering with him at the leaves. One adult male on a number of occasions hurried up when he saw a female grooming leaves and took over the grooming while the leaves were still in her hands, which he held along with the leaves.

The significance of this behaviour is obscure. Possibly it is some sort of redirected activity, but it may occur equally in chimpanzees who are grooming or being groomed by other individuals, and also in chimpanzees who are simply resting. It is rare for the behaviour to occur when an individual is completely on his own. Infants watch this behaviour closely. A one-year-old female watched as an adult male leaf-groomed. When he dropped his leaves, she picked one leaf and sucked it, then picked another and carefully scratched down it with one index finger. Infants of about one-and-a-half years have already begun to show adult leaf-grooming behaviour.

In this section I have discussed some of the behaviours which are almost certainly influenced to some extent by cultural traditions. Undoubtedly, given the investigative and manipulative tendencies of the young chimpanzee and his ability to learn through trial and error, almost all of the feeding and tool-using behaviours I have described could be 'invented' anew by each individual, especially since the behaviour of others in his group will serve to direct his attention to the relevant parts of the environment. However, in a species which is so well known for its imitative abilities it seems sensible to suppose that most, if not all, of the behaviours outlined above are not re-discovered by each chimpanzee, but are passed down from one generation to the next through observational learning in a social context.

Cultural Elements in the Sphere of Social Interactions

The extent to which cultural traditions may affect social behaviour in a chimpanzee society can never be fully appreciated until detailed analysis of the expressive gestures, postures, and calls of several different groups has been completed. Moreover, it will be of the utmost importance that the different investigators collaborate in standardizing terminology, data collection, and data analysis.

Table III lists the few differences in postures and gestures in different chimpanzee groups that I have been able to pick out of the existing literature. At best, however, this list does no more than point to a few aspects of behaviour that may differ in frequency or context in different groups; it does not reveal those subtle variations that, almost certainly, will result from the kinds of systematic comparative studies suggested. Moreover, for a variety of reasons, the differences listed may, in fact, be misleading.

In the chimpanzee, individuality (or personal idiosyncracies of behaviour) is a very pronounced characteristic, probably rivalled only by personality differences in man. Thus, in a given social context, each of four adult males may behave in a somewhat different and distinctive manner. Moreover, each male may display slightly different versions of an appropriate social response in reference to the particular individual with whom he is interacting. For example, when courting a certain female, male A may stare at her with his hair on end; male B may look away from her and shake branches; male C may perform a 'bipedal swagger', rocking upright from foot to foot; and male D may sit with shoulders raised and arms held slightly away from his body. Further, although male A courted this female by staring at her, he may shake branches at another female and achieve the same result in both cases.

This factor of individuality can significantly bias data (a) if one's data have been collected before the different individuals have been recognized; (b) if the data have been analyzed without reference to the individual performing a given behaviour; and (c) if data were collected over a short period of time when some individuals were seen frequently and others rarely. Let me cite an example from my own work at Gombe. My initial analysis of courtship behaviour described the different displays and listed the frequency with which each type was shown by adult males and by adolescent males. The results indicated that the bipedal swagger was a very typical display of adult males [van Lawick-Goodall, 1968]. In actual fact, two or three males were responsible for nearly all the bipedal swaggers I had observed in the sexual context; and it was typical only for them. This example shows how preliminary reports on social behaviour made by different investigators in different areas may be misleading and demonstrates that behaviours described as 'commonly seen' in a given group may, in fact, be typical behaviour for one or two individuals only.

Since it is not possible to compare, in any detailed and systematic way, the gestural and postural repertoire of the Gombe chimpanzees with that of other groups, I can offer here merely a few examples of the way in which observational learning and social contagion may function in the development of the

Table III. Some differences in social behaviour observed in different areas. These differences may reflect true cultural variations or merely idiosyncratic patterns in some individuals (see text)

Pattern	Context	Gombe	Kasakati	Uganda	Congo	Guinea	Other
Licking face	greeting	never seen	+ [3]				
Penis holding	greeting or reassurance	rare	+ [3]				
Bottom to bottom	reassurance	rare				+ [10]	
Mouth wide open; faces close but not touching	greeting	rare	+ [3]				
Reach back hand toward individual behind	mother signals to infant to climb dorsal	+	+ [3]	+ [5]	+ [7]	+ [10]	
	adult male inviting another to mount him during excitement	never seen			+[1]		
Attack 'victim' at close of charging display	general excitement	sometimes				[10] frequent	
Hunting for mammals	feeding	+	+[14–16]	+ [5]			Gambia[2]

References: [3] NISHIDA, 1970; [5] SUGIYAMA, 1969; [10] ALBRECHT and DUNNETT, 1971; [14] KAWABE, 1966; [15] NISHIDA, 1968; [16] SUZUKI, 1971.

1 Observed in unedited film material.
2 Free-ranging captive group (see footnote, table II).

communication patterns of the Gombe chimpanzees. These findings, in turn, will suggest the kinds of ways in which social modification may lead to cultural differences between separated groups of chimpanzees in the sphere of social interaction.

1. Male Charging Displays

Youngsters often watch, though from a discreet distance, when adult males perform charging displays. Subsequently, when the excitement is over, the youngster may execute a display of his own. A three-year-old male, for example, hurried to his mother's embrace as the alpha male of the community, Humphrey, gave pant-hoots preceding a charging display. The infant then watched as Humphrey charged across an open space, slapping the ground with his hands and stamping with his feet. He ended his display by jumping up and pounding with his feet on an empty 44 gallon drum at the observation area. When Humphrey had moved away, the infant left his mother, ran a short distance with much stamping of his feet on the ground, and then paused near the drum. After a moment he walked up to it, again paused, and then hit it gently twice with the knuckle of one hand. A three-year-old female also watched from the security of her mother's arms as an adult male displayed, charging along and stamping on the ground. When he had gone, she left her mother, walked to the place where he had displayed, and several times stamped with her feet on the ground.

The watching of and imitation of displays is by no means confined to infants. A male of about eight years of age was up in a tree when an adult displayed through the observation area past a group of chimps, dragged a large branch, and went from sight. A few moments later the youngster displayed; he followed almost precisely the same route and also dragged a branch.

In the above examples the locomotor patterns themselves appear to have been imitated. At other times it seems to be the context for the display that is learned by the youngster. Thus one juvenile male watched as several adult males charged across an open space, dragging and throwing branches, during very heavy rain. When they had gone, he also charged across the open space, first stamping his feet and then hurling himseld forward in a series of somersaults. Frequently, when a group of adult males arrive at a new food source or join up with a new group, they perform vigorous charging displays. If a juvenile or young adolescent male is with them, he is also likely to display, though usually keeping at a discreet distance from his elders as he does so.

One apparent difference between the charging displays of the Gombe chimpanzees and those observed by ALBRECHT and DUNNETT in Guinea may be a cultural variation. These investigators report that typically at the close of a charging display the performer briefly attacked some 'victim' [ALBRECHT and DUNNETT, 1971]. Although this kind of attack occurs at Gombe, it is relatively uncommon; it is not difficult to imagine that if a high ranking adult male established the pattern of pounding on another chimpanzee in this context, others might begin to follow suit.

2. Maternal Techniques

Many of the younger members of the chimpanzee society pay very close attention to infants leading, in some cases at least, to the imitation of particular maternal techniques. From the age of about four years, both male and female infants may direct some 'mothering' patterns towards younger individuals. These older infants, as well as juveniles and young adolescents, carry small babies around both in the dorsal and ventral positions, cuddle them, and groom them. Although the data have not yet been analyzed, it seems probable that juvenile and young adolescent females show more of this type of behaviour than do males of comparable age.

That some learning is involved in the acquisition of adequate maternal behaviour has been well demonstrated in captive chimpanzees [ROGERS and DAVENPORT, 1970]. But the extent to which maternal patterns may be imitated precisely has not yet been carefully studied. In this context the maternal behaviours of one old female at Gombe, Flo, and her offspring are of particular interest.

Flo's daughter Fifi was about six years old when her infant brother was born. Fifi showed much interest in the new baby, Flint; and when he was three months old, she was allowed to take him from his mother, play with him, groom him, and carry him. Some aspects of her behaviour in these respects were almost certainly the result of direct imitation of Flo's behaviour. Thus when Flo played with Flint, she very frequently lay on her back, dangled him above her with one of his wrists firmly clasped in her foot, and tickled him. This pattern was not seen in any other mother; but when Fifi played with Flint, she frequently held and tickled him in precisely the same manner.

Flo often played with Flint's penis when he was small, tickling it with either her hands or her lips. Other mothers show this behaviour very rarely. When Fifi was playing with Flint at that time, she very often played with his penis. Finally (although we have no quantitative data), Flo was observed to have a particular fondness for grooming the ears of her offspring. Often she held Flint down whilst he struggled to escape as she intently picked around one or the other of his ears. Fifi also frequently paid attention to Flint's ears when she was grooming him.

When Flint was about five months old, Flo initiated dorsal riding. On two different occasions about a week apart, we saw her push him up onto her back as she set off; both times he immediately slid down into the normal ventral position. Some two weeks later, Flint was first observed to travel for at least 40 yards on his mother's back. (The pair had been under very frequent observation every day during these months.) Later, on that same day, Fifi expended considerable effort in manoeuvring Flint until she had draped him over her shoulders. Flint was certainly not assisting his sister, and it is almost certain that Fifi was imitating her mother's behaviour.

Seven years later Fifi had her own first infant. During the first few months of his life (again, the data were not collected quantitatively) Fifi very often groomed her son's ears. Moreover, she repeatedly lay on her back and dangled Freud from one foot whilst she tickled him, a pattern which we still have not observed in other mothers.

After giving birth, Fifi continued to travel about frequently with her mother, who was still almost constantly accompanied by Flint. And it is of particular interest that Flint not only showed a great fascination for his nephew Freud, but also played with him, often by lying on his back and dangling the infant from one foot.

3. Significance of Individual Performance

At this point I should like to re-emphasize the potential significance of an individual's performance in the development of new cultural traditions in a society. Whilst we have not yet witnessed the diffusion of a new form of gestural or postural expression through the whole group, we have seen a novel behaviour imitated by a few other individuals. One example concerns 'wrist-shaking', in which the hand is shaken extremely rapidly to and fro. The gesture has been reported for a captive chimpanzee [GARDNER and GARDNER, 1969]. Not until 1964, at Gombe, was a chimpanzee observed to make this gesture; it suddenly appeared in a juvenile, Fifi (on whom we had made very regular observations throughout the previous year), when she was threatening an older female. A younger individual, Gilka, was with Fifi at the time. The following week Fifi was seen to repeat the gesture; the same week Gilka showed the pattern. Subsequently Gilka used the gesture very frequently indeed, in a variety of contexts; whereas Fifi also continued to wrist-shake, but infrequently and usually only in aggressive contexts. During the ensuing year the gesture was used by both individuals less and less often, and finally appeared to vanish from their repertoires.

Another example of the way in which social behaviour may be imitated occurred when a two-year-old infant, during play sessions, consistently sucked in her cheeks instead of showing the normal play-face. After a few weeks other infants with whom she frequently played also began sucking-in their cheeks during play sessions. The face itself was not novel, as it appears in most infants from time to time; the context in which it can be used, however, was new. Within the next few months the habit gradually disappeared.

The above are examples of behaviours which may become temporary 'fashions' amongst chimpanzees, particularly amongst play-mates. KÖHLER [1925] reported a whole variety of fashions of this sort which developed in his captive group, lasted for a few weeks or months, and then became obsolete. Whilst such temporary patterns do not represent cultural elements in the chimpanzee society, they deserve consideration because the very fact that unusual gestures or postures can suddenly appear and then be imitated in this way is significant. It suggests that an innovation of this sort might sometimes persist and gradually become incorporated into the repertoire of the entire community.

Probable Influence of Chimpanzee Social Organization on the Transmission of Culturally Influenced Elements of Behaviour

Japanese scientists have studied the ways in which new habits in Japanese monkeys pass from one individual to another and thus become traditional for the troop as a whole [e. g., ITANI, 1965; KAWAI, 1965; ITANI and NISHIMURA, this volume]. A good example comes from a troop living on the island of Koshima. Shortly after this troop had been provisioned for the first time with sweet potatoes, a juvenile female (nearly two years of age) began to wash the dirt from her potatoes by dipping them into water with one hand whilst she made cleaning movements with the other. This habit spread quite rapidly to other youngsters of similar age. It was also acquired by some adults who associated most closely with juveniles who had imitated the behaviour. It was apparent that kinship ties were particularly advantageous to the transmission of the innovation; mothers learned from infants, siblings from siblings, older offspring from their mothers. And, of course, once a mother had acquired the habit, she automatically passed it on to her own small infants. Five years after the invention of potato washing, 80% of the younger monkeys (under seven years) had adopted the habit. On the other hand, only 18% of the adult monkeys washed their potatoes; and these were all females who had close associations with youngsters. The remaining adults never learned the habit [KAWAI, 1965].

The habit of eating a new kind of food [ITANI, 1965] mentioned earlier in this chapter was acquired by a far larger percentage of adults, and far more rapidly, than the habit of washing food. This finding suggests that the adult is, on the whole, unlikely to acquire a new behaviour unless it results in an obvious reward.

Although we have not yet recorded the spread of a new habit through the Gombe Stream chimpanzee community, it is almost certain that, whilst novel behaviours would be most readily transmitted from one youngster to another and one family member to another, as in the monkeys, the unique structure of chimpanzee society would normally result in a very slow rate of transmission through the community as a whole. In a monkey group the infant, from the time when he has achieved a fair amount of locomotor independence, spends much time in a play group with other youngsters of similar age. Family ties are persistent, but the youngster associates with many non-family monkeys during the course of each day. The young chimpanzee, however, spends a great deal of his time in the company of his mother, together with his older sibling if he has one. Although his mother joins others from time to time, so that other infants are sometimes available for play, and although there are occasions when the infant will be part of a large mixed group, on the whole mothers, particularly as they get older, spend much time apart.

Figures 1 and 2 give some indication of the amount of time a mother may spend wandering about accompanied only by her own offspring. The data were obtained during four months of 'following' these mothers, each 'follow' lasting anything from an hour to a whole day. The total number of half-hour periods is given in each case. Figure 1a concerns a mother who has always been much inclined to spend long periods of time away from other chimpanzees; thus her two offspring seldom had the opportunity for play with similar aged youngsters during that four-month period. The family represented in figure 1b shows a rather more social mother whose family had greater opportunity to play with non-family members. This pattern may partly reflect the fact that her eldest offspring was a pre-adolescent male and may himself

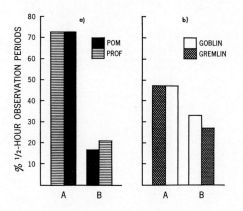

Fig. 1. a Percentage of half-hour observation periods (n = 339) of a chimpanzee family during four consecutive months, in which Pom (7-year-old female) and Professor Hamburg (1-year-old male) associated. A = With their mother only (no non-family members present); and B = with other groups of individuals that included possible play mates of similar age to themselves. *b* Similar data for another family of three in which the elder offspring, Goblin, was a 7-year-old male and the younger, Gremlin, a 2-year-old female. Based on 357 half-hour observations periods.

have initiated some of the contacts with other chimpanzee groups. Further analysis should clarify this point.

Figures 2a and b present similar data for the two youngest members of the Flo family, Flint and Freud. Flo had four offspring at the time, two adult sons, Faben and Figan, an adult daughter, Fifi, with her son, Freud, and Flint. At that time Figan still spent a good deal of time travelling with his mother, as did Fifi. Thus both Flint and Freud had frequent opportunity to play together; they also had the advantage of much association with an adult male, Figan. Taken together, the figures give some idea of the extent to which a chimpanzee, during his early impressionable years, has to rely on the companionship of his immediate family, an extent much greater than shown by any macaque.

This relative isolation means that whilst the chimpanzee infant is just as likely (or more likely) to invent new patterns as a young monkey, he has less chance of passing them on to his peers. The point should be made here that when young chimpanzees *have* the chance of frequently associating with other youngsters, novel behaviours are likely to be adopted by the group as a whole very rapidly. This statement has been documented for youngsters in captivity [e. g., KÖHLER, 1925]. Moreover, when the chimpanzees at Gombe were regularly fed with bananas and the youngsters met fairly often in quite large

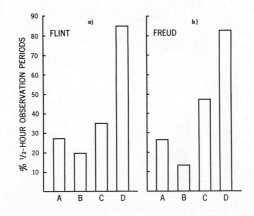

Fig. 2. Percentage of half-hour observation periods during the same consecutive four-month period in which the 2 youngest members of the Flo family associated with others. *a* Flint's associations (based on 231 half-hour periods): A = % total number of ½-hour periods Flint spent alone with his mother (Flo); B = % time Flint had opportunity to play with youngsters other than nephew Freud; C = % time Flint alone with mother and sister Fifi; D = % time Flint with one or more family members, when no non-family present. *b* Freud's associations (based on 225 half-hour periods): A = % total time Freud alone with mother (Fifi); B = % time Freud had opportunity to play with youngsters other than uncle Flint; C = % time with Fifi, Flo, Flint and (sometimes) older uncle (Figan); D = % time with one or more family members (no non-family present).

groups, a novel pattern (the sucked-in-cheeks face in play) spread rapidly from one infant to another. Normally, however, wild chimpanzee youngsters have little opportunity for play in groups so that when they invent new behaviours, they may spread to only a few youngsters over a long period of time. Also, since adults are unlikely to imitate behaviours that have no obvious reward, social innovations (such as using a twig to 'inspect' the genital area of a female) are unlikely to be acquired by older family members.

Now let us look at the other side of the picture. The fact that the chimpanzee infant associates for much of the time only with his immediate family means also that his mother and older siblings will be the most frequently available models for his imitative behaviour. Although our data have not yet been analyzed with sufficient detail to determine the extent to which subtle behavioural characteristics may be shared by the members of a family, we have evidence of some family traditions within the Gombe community.

In most primate species travel routes can be defined as traditional for the group as a whole. In chimpanzee societies, on the other hand, individuals

Fig. 3 Adult female kissing adult male in greeting at Gombe Stream.

Fig. 4 Charging display of a Gombe male.

*Fig.*5 Sequence showing family group. *a* The second eldest son, adult male Figan,

sometimes show strong preferences for following one path rather than another when they travel between given points. A youngster follows his mother during the years of his dependency and becomes familiar with her preferred routes. It is not surprising that he may continue to use many of them once he becomes independent. Whilst I have been working on this chapter, chimpanzees have passed my house on the lake shore several times. On each occasion one or more members of the Flo family were present, and in each case it was a family member who led the group. And I remember, back in 1964, following Flo and two of her youngsters along this same route. Analysis of the data will undoubtedly reveal similar traditional paths for other families within the community and also, almost certainly, traditional nesting places.

In some instances the food preferences of a mother may be reflected by her offspring. Hunting is typically a male activity at Gombe, although females have been known to kill on rare occasions. In one family, however, on three occasions the mother has been seen making hunting attempts herself; and twice her two daughters (juvenile and adolescent) joined in [A. PUSEY and M. HANKEY, personal communication]. Another time the family of three were seen feeding on a monkey; since no other chimpanzees were present, it seems likely that the mother had made the kill herself. It is possible that the hunting behaviour of this female, if it is maintained by both her daughters, might eventually have a considerable influence on the role of females in hunting at Gombe.

I have already described how the female, Fifi, imitated certain maternal techniques that were typical for her mother. The fact that her own younger brother showed one of these behaviours when playing with infants (dangling the baby from one foot during play) suggests that he may have imitated his sister. On the other hand, he may have behaved thus as a result of being played with in that way when he was small.

One other social behaviour pattern has been passed from mother to daughter in a different family. The mother, Madam Bee, was victim of a paralytic disease during which she lost the use of one arm. At the time her two daughters were about eight and one and one-half years old, respectively. As a result of her affliction, Madam Bee has developed an unusual social grooming technique. Normally when a chimpanzee grooms a companion, he parts the hair with one hand; and with the other, and often with lips as well, picks at

---

approaches his old mother, Flo, who is with her daughter and grandson, and her 8-year-old son Flint. *b* Subsequently the other members of the family all groom Figan as part of their greeting.

small flakes of dried skin, and so on. When a chimpanzee is grooming his own arm he usually first scratches down the limb quite vigorously before proceeding to part the hair and pick out particles with his hand and lips. Madam Bee very often shows the vigorous scratching on her companion's body that is typical of self-grooming. Today her eldest daughter shows normal social grooming techniques; but the younger one, now about seven years old, often scratches her mother vigorously before grooming her [M. HANKEY and A. PUSEY, personal communication].

There is one other aspect of the chimpanzee social structure which we should consider here. When the young male first starts moving independently of his mother, he frequently associates with mature males; and whilst he may occupy a peripheral position in such a group, he nevertheless watches their behaviour very closely. Adolescence is a time of major hormonal and behavioural changes, and it is not unreasonable to suppose that the young male may imitate some of the behaviour of the adult males he watches so closely. Indeed, I have already cited examples showing that this imitation is indeed so with regard to charging displays. Subsequent study will determine whether the adolescent makes precise imitations of some of the individual patterns characterizing particular adult males. If this were so, it would provide an important pathway for the transmission of such behaviours, since the adolescent still associates frequently with his family at this time and the patterns might well be imitated by a younger sibling.

So far we have discussed ways in which cultural elements may be transmitted from one individual to another within a chimpanzee community. We should also consider what mechanisms are available for the transmission of a behaviour from one community to another. ITANI [1965] comments on the fact that it is difficult for new habits to be transmitted from one troop of Japanese monkeys to another. In this species, as in most non human primates, it is young males who normally transfer to new troops. And young male Japanese monkeys, whether they have remained in their natal troops or transferred to new ones, occupy a peripheral position. We have already seen that new behaviours are only transmitted readily between monkeys who associate closely, so that, even if a behaviour is imitated by another peripheral male, it is unlikely to be acquired by monkeys in the central part of the troop.

In the chimpanzee there is increasing evidence of the transfer of young females between communities. At Gombe young females in oestrus frequently visit neighbouring communities, where they associate with and mate with the host males. Normally they return to the natal group at the end of oestrus, but two young females have joined our community from elsewhere and stayed on.

Fig.6 Flo dangling Flint during play in her characteristic manner.

NISHIDA and KAWANAKA [1972] have now recorded 39 cases of females transferring from one unit-group (community) to another.

It follows that, once the transfer female has started a family, any aspects
of her behaviour which are novel to individuals of her new community may be
transmitted to them probably in the same way and at the same rate as would a
new trait in her natal community. This process, as we have seen, is likely to be
slow. However, provided she is successful in raising a family, the behaviour
will probably be kept alive; and if it is an adaptive one, it may ultimately be
incorporated into the repertoire of the community as a whole.

Cultural Elements in Chimpanzee Groups in Captivity

The behaviour of chimpanzees in captivity has given us much insight into
their capacity for learning new elements of behaviour. Home-raised chim-

panzees have imitated all manner of performances from washing the dishes to putting on lipstick before the mirror with pursed lips [e.g., Kohts, 1935; Kellogg and Kellogg, 1933; Hayes and Hayes, 1952; Gardner and Gardner, 1969]. Yerkes discusses the way in which newly acquired adaptive behaviour (such as the ability to operate a push-button fountain for drinking) was transmitted through the colony at Orange Park. Once such a behaviour had been learned by one individual it was 'passed along from individual to individual by imitative process and from one generation to the next by social tradition'. Other novel behaviours acquired by these chimpanzees, such as spitting through the teeth, squirting water, clapping hands, and using objects (such as balls, keys, hammers, and so on) were learned by observing humans and became cultural traditions in the chimpanzee colony [Yerkes, 1943].

Today there are a number of social groups of chimpanzees in captivity in environments that, although they cannot be compared with that of the natural habitat, are nevertheless a great improvement over the conventional zoo or laboratory cage. Close observation of these groups will provide unique information with regard to the transmission through the generations of cultural elements in the sphere of social interaction. Two of these groups are at present being studied by researchers using methods of data collection similar to those in use at the Gombe Stream Research Centre[4]. One of these groups, which was formed by Menzel at the Delta Regional Primate Research Center and extensively studied by him [Menzel, 1971, 1972, this volume], is the nucleus for the new Outdoor Primate Facility at Stanford (Gombe West). The seven adolescent and young adult animals comprising this group were wild-born and have lived together for years. The other group of 12 adult chimpanzees is the property of Lion Country Safaris in Florida. These individuals were introduced to each other as adults and come from extremely diverse backgrounds.

Although none of the data collected on these captive groups has been systematically compared with that collected on the wild chimpanzee community at Gombe, several differences in social behaviour patterns between the captive and the wild groups are already clear. For example, playful behaviour between adults and between adolescents is very, very much more frequent in both captive groups. Moreover, patterns used to initiate play in the captive groups are often those seen typically during aggressive encounters at Gombe. The bipedal swagger, for instance, which occurs principally as an aggressive

4   The Florida group was intensively studied for two months by C. Gale and W. Cool; the Stanford group by W. McGrew, P. Midgett, P. McGinnis, C. Tutin and other students.

Fig. 7 Four-year-old male using a stick to club an insect on the ground.

display in the wild, is a common invitation to play in the Florida and Stanford groups. At the same time a pattern seen typically in play situations at Gombe, the somersault, was for a while incorporated into his aggressive displays by one individual in the Stanford group. At Gombe, as we have seen, each chimpanzee (with the exception of dependent youngsters) sleeps in his own nest at night. In both of the captive groups the chimpanzees sleep curled up in small groups, either in sleeping hutches (Stanford group) or big communal hay nests on the ground (Florida group).

Two behaviours are peculiar to the group at Florida. Whilst the chimpanzees at Gombe sometimes leave their nests to feed on moonlit nights and, very rarely, may travel at such times, midnight activity is *normal* for the chimpanzees at Florida. Not only do they frequently move about at night; but they may play, groom, and perform charging displays. One tremendous battle for status took place during the pitch darkness of a moonless night. At Gombe, the most common submissive posture in females, and probably also in juvenile and young adolescent males, is presenting, or turning the rump towards the

8a

Fig.8. a and *b* Flo, with Flint dorsal, approaches adult male in greeting. The mother shows characteristic pant-grunting; her infant shows exactly the same behaviour.

dominant animal. This posture was only seen on *very* rare occasions during two months at Florida, even in the sexual context.

One adolescent female (Belle) in the Stanford group began to use twigs as tools to poke at and even remove her loose deciduous teeth. This habit spread through the group; dental 'grooming' of this sort has never been observed at Gombe. Eventually, Belle began to use tools to groom the teeth of her companions and actually performed an extraction on a young male [McGrew and Tutin, 1972, in press].

Earlier I discussed the different ways in which an infant chimpanzee

8b

learned typical chimpanzee behaviours. We should expect that youngsters growing up in abnormal conditions would develop abnormal behaviours, a situation that indeed occurs. In the Stanford group, for instance, one young male typically initiates copulation with a display that is undoubtedly peculiar to himself [C. TUTIN, personal communication]. Another female taps strange tattoos with her fingers. And in Florida one female successfully gave birth to an infant and subsequently showed some abnormal mothering techniques. She cared for the infant adequately in that she fed him and frequently carried him around, but she also very often detached him and laid him beside her on the ground. From the time when he was a few days old she often moved away short distances, leaving him behind. On these occasions she occasionally buried him under a pile of hay. It is interesting that another captive female, who showed good mothering behaviour on the whole, also left her infant behind very frequently whilst she moved about the cage. This mother usually constructed a large nest of straw on the ground in which to deposit her baby; if he whimpered, she rushed back at once [M. BADHAM, personal communication].

Fig. 9. Juvenile termite-fishing.

Will other individuals, particularly infants born into the groups, imitate any of these abnormal patterns? Are some of these unusual social behaviours more likely than others to be acquired by developing youngsters? Answers to these and other questions should significantly forward our understanding of the role played by cultural elements in chimpanzee social behaviour. We shall begin to learn which items of behaviour are the most flexible and which are the least susceptible to change and thus shed new light on problems of adaptive modification in the chimpanzee.

Discussion

In this chapter I have presented an over-simplified account of some of the more obvious elements of behaviour in chimpanzee society that have probably been culturally acquired. Much of the material is speculative, but will, I hope, serve to draw attention to areas that with increasing sophistication of methodology and increasing cooperation between investigators could lead to major advances in the understanding of an extremely complex subject.

The study of the chimpanzee, both in the field and in the laboratory, has revealed a creature who is almost startlingly close to man in many aspects of his behaviour. For these apes show a great many characteristics once thought to be uniquely human. They show sophisticated cooperation in some situations such as hunting and in specially devised laboratory experiments [CRAWFORD, 1937]. They are the only primates other than man known to share food among adults in the wild. Insightful problem-solving, or ideation, is well documented by laboratory experiments [e.g., KÖHLER, 1925; YERKES and YERKES, 1929] and also occurs in the wild [e.g., VAN LAWICK-GOODALL, 1970]. Recent laboratory research has revealed that chimpanzees have the ability to make symbolic generalizations [GARDNER and GARDNER, 1969; PREMACK, 1971] and to make associations across sensory modalities [DAVENPORT and ROGERS, 1970]. Moreover, they show behaviours that might perhaps be precursors to the human qualities of altruism, art, and moral values. Some preliminary thoughts on these aspects of chimpanzee behaviour will be presented elsewhere [VAN LAWICK-GOODALL and HAMBURG, in preparation].

Why, in view of this degree of sophistication, does the chimpanzee in nature not show a *more* advanced level of cultural behaviour? We must realize that, despite the complexity of his expressive signals, the chimpanzee has no method of communication that can be compared with human language. Certainly, given the intellectual comprehension, extremely complex behaviours can be acquired through observation and imitation. But without a spoken language (or a written one or a very sophisticated sign language), it is virtually impossible to convey precise information about the past or precise and detailed plans regarding anything but the immediate future. Above all, it is impossible to start on the elaboration of conceptual ideas that has led to the evolution of our uniquely human intellect, sensitivity, altruism, aesthetic appreciation, religious belief, and ability to love. It seems very likely that the development of language opened up the evolutionary road leading to the diversity and richness of culture in men.

Language provides a unique pathway for the transmission of cultural behaviour. Man no longer needs to rely on the physical presence of a model in order to acquire a new element of behaviour. He can, if he wishes, model his behaviour on a person who lived a thousand years ago in a distant land or on a being who has no existence but was merely the creation of another's mind. Nevertheless, whilst we cannot make direct comparisons between chimpanzee and human 'culture', as such, a careful examination of culturally influenced elements in a chimpanzee society and of the way in which they are transmitted through the group and from generation to generation may help to illuminate some of the evolutionary mechanisms which have played their role in creating man as we know him today.

Acknowledgements

I should like to express gratitude to Tanzania's President, Mwalimu Julius Nyerere, to the Prime Minister, and to the many other Government officials who have encouraged and assisted us in our work. The research was initially financed by the Wilkie Foundation and subsequently by the National Geographic Society, the Science Research Council, and the Wenner Gren Foundation; I should like to thank all these organizations. At present the work is financed by the Grant Foundation; and I am particularly grateful to Dr. Douglas Bond, the President, and Mr. Philip Sapir, the Director.

I could not have carried out the work in the field without the help of Hugo van Lawick; I should like to thank him and also the many students and Tanzanian research and field assistants who have helped us collect an ever-increasing volume of information on chimpanzee behaviour. For allowing me access to unpublished observations for this paper, I thank R. Wrangham, F. Plooy, A. Pusey and M. Hankey (Gombe information): C. Gale and W. Cool (Lion Country Safari, Florida information) and M. Badham. I am also grateful for the detailed information sent by E. Brewer and S. Brewer from the Gambia.

The Gombe Stream Research Centre is now formally affiliated with Stanford University and has enduring links with the Universities of Dar es Salaam and Cambridge; I should like to thank Professor Hamburg, Professor Msangi, and Professor Hinde of these institutions for their help and encouragement. Finally, I express my gratitude to the late Dr. L.S.B. Leakey, who gave me the opportunity to start the research in the first place.

References

Albrecht, H. and Dunnett, S.C.: Chimpanzees in western Africa (Piper, München 1971).

Beatty, H.: A note on the behavior of the chimpanzee. J. mammal. *32:* 118 (1951).

Bernstein, I.S.: Response to nesting materials of wild-born and captive-born chimpanzees. Anim. Behav. *10:* 1–16 (1962).

Crawford, M.P.: The co-operative solving of problems by young chimpanzees. Comp. psychol. Monogr., vol. 14/2 (The Johns Hopkins Press, Baltimore 1937).

Darby, C.I. and Riopelle, A.J.: Observational learning in the rhesus monkey. J. comp. physiol. Psychol. *92:* 94–98 (1959).

Davenport, R.K. and Rogers, C.M.: Differential rearing of the chimpanzee; in Bourne The chimpanzee, vol. 3, pp. 337–360 (Karger, Basel 1970).

Eibl-Eibesfeldt, I.: Ethology (Holt, Rinehart & Winston, New York 1970).

Gardner, R.A. and Gardner, B.T.: Teaching sign language to a chimpanzee. Science *165:* 664–672 (1969).

Hayes, K.J. and Hayes, C.: Imitation in a home-raised chimpanzee. J. comp. physiol. Psychol. *45:* 450–459 (1952).

ITANI, J.: On the acquisition and propagation of a new food habit in the troop of Japanese monkeys at Takasakiyama; in IMANISHI and ALTMANN Japanese monkeys: a collection of translations, pp. 52–65 (University of Alberta Press, Edmonton 1965).

ITANI, J. and SUZUKI, A.: The social unit of chimpanzees. Primates *8:* 355–381 (1967).

IZAWA, K. and ITANI, J.: Chimpanzees in Kasakati basin, Tanganyika. Kyoto Univ. Afr. Stud. *1:* 1–255 (1966).

KAWABE, M.: One observed case of hunting behavior among wild chimpanzees living in the savanna woodland of western Tanzania. Primates *7:* 393–396 (1966).

KAWAI, M.: On the system of social ranks in a natural troop of Japanese monkeys: (1) basic rank and dependent rank; in IMANISHI and ALTMANN Japanese monkeys: a collection of translations, pp. 66–86 (University of Alberta Press, Edmonton 1965).

KELLOGG, W. N. and KELLOGG, L. A.: The ape and the child: a study of environmental influence upon early behavior (McGraw Hill, New York 1933).

KÖHLER, W.: The mentality of apes (Harcourt & Brace, New York 1925).

KOHTS, N.: Infant ape and human child. Sci. Mem. Mus. Darwin, Moscow *3:* 1–596 (1935).

KORTLANDT, A.: Chimpanzees in the wild. Sci. Amer. *206:* 128–138 (1962).

KORTLANDT, A.: Bipedal armed fighting in chimpanzees. Proc. 16th Int. Congr. Zoo. *3:* 64 (1963).

KORTLANDT, A.: How do chimpanzees use weapons when fighting leopards? in Yr. Bk. Amer. Phil. Soc., pp. 327–332 (Philadelphia 1965).

KORTLANDT, A.: Handgebrauch bei freilebenden Schimpansen; in RENSCH Handgebrauch und Verständigung bei Affen und Frühmenschen, pp. 59–102 (Huber, Bern 1968).

KUMMER, H.: Primate societies (Aldine-Atherton, Chicago 1971).

LAWICK-GOODALL, J. VAN: The behaviour of free-living chimpanzees in the Gombe Stream Reserve. Anim. Behav. Monogr., vol. 1/3 (Balliére, Tindall & Cassell, London 1968).

LAWICK-GOODALL, J. VAN: Tool-using in primates and other vertebrates; in LEHRMAN, HINDE and SHAW Advances in the study of behavior, vol. 3, pp. 195–249 (Academic Press, New York 1970).

LAWICK-GOODALL, J. VAN: The behavior of chimpanzees in their natural habitat. Amer. J. Psychiat. *130:* 1–12 (1973).

MCGREW, W. C. and TUTIN, C. E. G.: Chimpanzee dentistry. J. amer. dent. Ass. *85:* 1198–1204 (1972).

MCGREW, W. C. and TUTIN, C. E. G.: Chimpanzee tool-use in dental grooming. Nature, Lond. (in press).

MENZEL, E. W.: Patterns of responsiveness in chimpanzees reared through infancy under conditions of environmental restriction. Psychol. Forsch. *27:* 337–365 (1964).

MENZEL, E. W.: Communication about the environment in a group of young chimpanzees. Folia primat. *15:* 220–232 (1971).

MENZEL, E. W.: Spontaneous invention of ladders in a group of young chimpanzees. Folia primat. *17:* 87–106 (1972).

MERFIELD, F. G. and MILLER, H.: Gorillas were my neighbours (Longmans, London 1956).

NAPIER, J. R.: Prehensility and opposability in the hands of primates. Symp. zool. Soc., Lond. *5:* 115–132 (1961).

NISHIDA, T.: The social group of wild chimpanzees in the Mahali mountains. Primates *9:* 167–224 (1968).

NISHIDA, T.: Social behavior and relationship among wild chimpanzees of the Mahali mountains. Primates *11:* 47–87 (1970).

NISHIDA, T. and KAWANAKA, K.: Inter-unit-group relationships among wild chimpanzees of the Mahali mountains. Kyoto University Afr. Stud. *VIII:* 131–169 (1972).

NISSEN, H.W.: A field study of the chimpanzee. Comp. psychol. Monogr., vol. 8/1 (The Johns Hopkins Press, Baltimore 1931).

PREMACK, D.: Language in chimpanzee? Science *172:* 808–822 (1971).

REYNOLDS, V. and REYNOLDS, F.: Chimpanzees of the Budongo forest; in DEVORE Primate behavior, pp. 368–424 (Holt, Rinehart & Winston, New York 1965).

ROGERS, C.M. and DAVENPORT, R.K.: Chimpanzee maternal behavior; in BOURNE The chimpanzee, vol. 3, pp. 361–368 (Karger, Basel 1970).

SAVAGE, T.S. and WYMAN, J.: Observations on the external characters and habits of the *Troglodytes niger,* Geoff... and on its organization. Boston J. natur. Hist. *4:* 377–386 (1843/44).

SUGIYAMA, Y.: Social behaviour of chimpanzees in the Budongo forest, Uganda. Primates *10:* 197–225 (1969).

SUZUKI, A.: On the insect eating habits among wild chimpanzees living in the savanna woodland of western Tanzania. Primates *7:* 481–487 (1966).

SUZUKI, A.: Carnivority and cannibalism observed among forest-living chimpanzees. J. Anthrop. Soc. Nippon. *79:* 30–48 (1971).

TYLOR, E.B.: Primitive culture: Researches into the development of mythology, philosophy, religion, language, art, and custom (Murray, London 1871).

YERKES, R.M.: Chimpanzees: a laboratory colony (Yale University Press, New Haven 1943).

YERKES, R.M. and YERKES, A.W.: The great apes: a study of anthropoid life (Yale University Press, New Haven 1929).

Author's address: Dr. JANE VAN LAWICK-GOODALL, Gombe Stream Research Centre, P.O. Box 185, Kigoma, *Tanzania* (East Africa)

Symp. IVth Int. Congr. Primat., vol. 1: Precultural Primate Behavior,
pp. 185–191 (Karger, Basel 1973)

Implications of a Primate Early Rearing Experiment for the Concept of Culture[1]

C. M. ROGERS[2]

University of Guelph, Guelph, Ontario

A way of assessing the extent to which the stable day-to-day behavior of gregarious animals, particularly their interpersonal interactions and communications, is, in fact, culturally influenced or dependent on social learning is to see how an animal would behave in social situations after having been deprived of all opportunities for social learning in early life. Following this approach, I will present some general qualitative conclusions drawn by myself and colleagues from extended (15 years) longitudinal observations of the social behavior of two groups of chimpanzees at the Yerkes Regional Primate Research Center. The control (wildborn) group consisted of five female and three male chimpanzees that spent approximately the first year of their lives in the wild, in the care of their mothers, and were then captured, brought to the laboratory, and caged in age-peer pairs and in groups. The experimental (restricted) group was composed of animals that were removed from their mothers within a few hours of birth and reared for two years in various drastically impoverished and extremely monotonous environments, which excluded any maternal contact, prevented any visual stimulation from the environment outside the rearing box, and, for most of them, prevented any social contact [DAVENPORT and ROGERS, 1970]. In the restricted group there were also five females and three males who matched the wildborns in age. Since about five years ago, when the animals had all become young adults (at 13–15 yr.), the two groups have been housed together in a compound.

When the restricted animals were first placed in social situations, at approximately three years of age, the contrast between them and wildborn

1   This research was supported in part by the Ford Foundation and by NIH grant No. M-1005, H-5691, HDO-3720, and FR-00165 to Yerkes Regional Primate Research Center.
2   Present address: Auburn University, Auburn, Alabama.

age-peers was striking. Social interactions in groups of restricted animals were rare. Individuals separated themselves from other animals and engaged in such self-directed activities as solitary play and stereotyped behavior to the exclusion of interaction with other animals.

All of our restricted animals developed idiosyncratic patterns of stereotyped behavior in early infancy. These behaviors are characterized by the frequent, almost mechanical, repetition of movements that vary only slightly in form from time to time and, to the human observer, serve no obvious function. There are three general types: (1) rocking, turning, swaying movements involving the whole or a major portion of the body; (2) thumb-sucking, eye-poking, and other repetitive movements of individual body parts; and (3) posturing [DAVENPORT and MENZEL, 1963]. It may well be that the compulsive repetition of these behaviors serves to insulate the animal and may even provide comfort, since frequencies rise during periods of stress. The early practice of these self-directed behaviors was so frequent and such a pervasive part of the behavioral repertoire of these animals that it may have interfered with their learning of social behaviors for several years during which they were kept in groups of their own kind, where they had only occasional contact with wildborn animals. When these restricted chimpanzees were from seven to nine years old (approaching, or just at, puberty), their exposure first to paired and then to group living with wildborn age-peers was increased gradually until the present permanent living condition was established. Their initial pervasive fearfulness gradually subsided over the subsequent years of group living, and interactions both with each other and with wildborns have become more frequent. First, fragmented and distorted approximations of various social interaction patterns began to appear. With time and experience, these tended to become more complete and to approximate more closely sequences exhibited by their wildborn age-peers.

At first the restricted animals did not perform in a recognizable manner such social patterns as grooming, sexual behavior, aggression, and aggression avoidance. Additionally, and of at least equal importance, they did not respond appropriately to social signals. These observations are, we believe, a clear demonstration of the importance of learning in the acquisition of a set of species-typical patterns that are quite stylized and that are exhibited by the members of all normal chimpanzee groups.

Over the years our previously socially-inept, restricted subjects have learned some of the patterns of social interaction, including the appropriate giving of and responding to social signals. This learning has occurred in spite of the fact that they still practice a variety of stereotyped behaviors at high levels for long periods of time, particularly when there is little social activity in the compound. But they interrupt these behaviors to engage in social interactions, and they sometimes continue to stereotype while interacting socially.

The gradual improvement in both quality and frequency of social interactions has taken a long time; their approximations are still quite imperfect. We do not know the extent to which they will ultimately improve, but our restricted subjects have advanced to a point where their patterns and signals now elicit appropriate responses from the wildborns. Now that the restricted chimpanzees respond to social signals more appropriately, frequencies of aggression by wildborn toward restricted chimpanzees have declined. Initially, any approach by a wildborn provoked cowering, flight, or falling prone and screaming; and these behaviors frequently inspired wildborn aggression. Ordinarily, if a more or less casual approach of another animal is ignored or produces little more than a glance and a pause in ongoing activity, no aggression by the approaching animal follows. Most of the restricted chimpanzees have finally learned this lesson.

Another example related to this one is behavior in response to aggressive signals. The wildborn males occasionally threaten other animals or give a display which does not seem to be directed toward a specific individual. The display is quite similar to a threat, except that in a group of normal animals it does not usually result in an attack or charge directed toward a specific individual. Usually, when a male displays, other wildborn animals either move out of his way, with little or no outward sign of fear or excitement, or appear to ignore him. This response is quite different from the flight, screams, and attempts to placate and to elicit help exhibited when they are directly threatened and charged. At first, the restricted animals did not distinguish threats directed toward them from threats directed toward others or from display behaviors. When any of these events occurred in their vicinity, their typical response was to fall to a prone position while screaming loudly. Frequently, the displaying male would literally run over them; sometimes he would attack them, and occasionally he would shift his attack from another animal to them. Gradually, the restricted animals have become more selective in their reactions; and, although they frequently appear tense when such events as displays occur, they have learned not to attract attention to themselves. Now, when they are charged, they seldom drop to the ground but flee as do the wildborns.

The grooming sequence, usually a dyadic pattern, is a complex social interaction. As one animal grooms, the other becomes inactive, immobile, and pliable. If the groomer pauses, the recipient may reposition himself and thereby induce the resumption of grooming; or the roles may be reversed with the formerly active animal becoming the compliant recipient. Early in the social groupings, restricted animals tended to respond to grooming overtures by such totally inappropriate behaviors as tenseness, avoidance, whimpering, even screaming. Gradually they have learned to differentiate the usual grooming overtures from threat patterns and, finally, have become cooperative partners. Conversely, as their own grooming attempts have progressed from random pokings and thumpings to closer (though still far from perfect) approximations of grooming, wildborns have become more tolerant of their overtures and respond less often with avoidance, threats, or slaps.

Copulation is another complex social interaction. Either a male or a female may solicit copulatory behavior by giving the appropriate set of signals to a member of the opposite sex. Even though apparently well executed, these signals occasionally do not produce the appropriate response. But the closer they approximate the species-typical postures and gestures, the greater the likelihood that the next step in the pattern will occur. The sophisticated male will generally confine his sexual solicitations to estrous females with conspicuously swollen perianal regions. The restricted males at first exhibited elements of sexual solicitation behavior indiscriminately toward estrous females, anestrous females, and other males. Frequencies of inappropriate solicitations have gradually declined.

Sophisticated animals of either sex may solicit. The female may repeatedly place herself in front of her choice and pause, with her knees moderately flexed, to display her swollen perianal region. On occasion she may even back toward the male if she sees that he has a penile erection. None of our restricted females exhibited any of these signals early in the group association and even now their positionings before the male are only occasional and characteristically lack the leg flexion. Soliciting by the normal male is highly stylized and involves squatting with knees spread wide to display an erect penis; most wildborn males accompany this by slapping the ground with open palms. If a female does not present to him, he may after several seconds rise to an erect posture and execute a short dance in some respects similar to a threat display. He will then frequently alternate from one pattern to the other if not interrupted by a sexually-presenting female. Restricted males have consistently failed to give more than very vague approximations of these patterns. One male's attempts in this direction were frequently misinterpreted by chimpanzee females as threat posturings, until a few daring wildborn females eventually responded by backing up to him, with the result that copulation ensued. Early in his social education this male also attempted to copulate in inappropriate situations and ways, for example, covering and thrusting against anestrous females and other males, covering and thrusting against reclining females, orienting inappropriately, and occasionally thrusting against another animal's side or head. Correcting such errors is not merely a matter of the practice of various postures and mechanical movements but requires an awareness of, and adjustment to, movements and signals of the other animal.

Early in the group association, all of the sexual activity of restricted animals occurred exclusively with wildborn partners; and successful intromissions of restricted males were first performed with sophisticated, older, wildborn females. These females literally backed the male into a corner and effected insertion by proper positionings in relation to the male [ROGERS and DAVENPORT, 1969].

We feel that all of these observations indicate that species-typical patterns of social interaction and communication, while demonstrably characteristic of wild chimpanzee groups, are not inborn but must be acquired through social experience (learning). They are phenomena which should, therefore, be considered cultural.

A further point should be made about the effect of social learning on sexual behavior. In a previous paper [ROGERS and DAVENPORT, 1969] we noted that the sexual behavior of our restricted-reared animals was less severely damaged than was that of a group of chimpanzees reared by humans in the Yerkes ape nursery [RIESEN, 1971]. We suggested that the substitution of the human maternal care provided these animals with inappropriate experiences during infancy that later seriously affected their ability to respond to the appropriate social cues and signals related to sexual interaction. It should be noted that, rather than exhibiting abnormal attempts at copulation, most of these nursery-raised animals showed no inclination to copulate, no apparent sexual arousal. These observations differ from the behavioral abnormalities we have described. They emphasize the fact that inappropriate social learning can be at least as damaging to the behavioral development of the organism as the lack of opportunity for learning [HEDIGER, 1950].

We have reported elsewhere [ROGERS and DAVENPORT, 1970] that there was a marked difference in the quality of maternal behavior of primiparous mothers, depending on whether they had themselves had prolonged mothering. Females that had spent less than 18 months with their mothers in infancy were, in general, poor mothers at first; and most of their infants would not have survived if they had not been removed and cared for by human beings. These mothers did not harm their infants, but they did not care for or nurse them.

At the time we reported these differences we offered an hypothesis about how nursing first takes place. The nursing interaction is not automatic, and a good deal of learning is required for skillful performance by the mother. Chimpanzees do not normally carry the infant near the breast, and the newborn is not physically capable of pulling itself to the breast. But both mother and infant have a set of more or less automatic responses that can lead to initial nursings. The infant whimpers, squeals, squirms, and kicks when it is in discomfort. The mother tends to be distressed by these signs of discomfort from the newborn. This distress produces a variety of maternal responses, among which is repositioning of the infant on her body. As a result of this initially more or less random repositioning, the infant eventually reaches a point from which it can contact the nipple. Repetitions of the process produce rapid learning of the component behaviors by the mother. The reinforcement for the mother is, we think, a reduction of her tension brought about by a reduction in distress signals from the infant. With repeated occurrences of the nursing sequence, the physiological drives of hunger (in the infant) and breast tension (in the mother) play an increasingly important role in motivating nursing. This is social learning of a relatively basic form.

In conclusion, culture in the sense of established, transmitted behavioral patterns of a group, plays a definable role in the development of individual animal behavior. We have seen our restricted chimpanzees develop from fearful, withdrawn animals who performed stereotyped movements virtually all of the time into animals who are at least approaching social normality as a result of living in a group with wildborn animals. This development we consider an effect of social learning. But the possibility that social patterns evolve toward species-typical behavior because of an innate tendency of members of a species to perform elements of certain patterns in certain ways should not be discounted [Mason et al., 1968]. Some elements of the patterns appeared in the restricted animals before social exposure to the group as a whole; but they were not coordinated, not in proper sequence, and frequently not performed in appropriate contexts. The learned portion of the behaviors appears to be the proper organization of the innate elements of behavior and the recognition of the intent of other animals, that is, the meaning of social signals.

There is a final point that should be made about culture and social learning. Although we tend to think of culture as enriching experience which increases the potential of the individual to develop his own aptitudes, capacities, proclivities, and individual talents, there are other opposite effects. These operate to exert a leveling effect on the developing individual and function to reduce individual variability, thereby producing a conforming individual who communicates and responds in highly stereotyped and predictable ways. At the outset of our study we fully expected that our restricted-reared subjects, deprived of the rich experiences of a varied environment in which there are stimuli constantly impinging on the animal and a multitude of opportunities to respond in a variety of ways to these inputs, would develop about as much variability as peas in a pod. This was clearly not the case. These animals exhibited striking individual differences in their responses in most of our evaluation procedures and in their early social contacts. Such response variability is obviously incompatible with efficient social communication. A large part of their eventual socialization was the elimination of idiosyncratic behaviors and the learning of patterns more typical of the species.

References

DAVENPORT, R.K. and MENZEL, E.W.: Stereotyped behavior of the infant chimpanzee. Arch. gen. Psychiat. *8:* 99–104 (1963).

DAVENPORT, R.K. and ROGERS, C.M.: Differential rearing of the chimpanzee; in BOURNE The chimpanzee, vol. 3, pp. 337–360 (Karger, Basel 1970).

HEDIGER, H.: Wild animals in captivity. (Butterworths, London 1950).

MASON, W.A.; DAVENPORT, R.K., and MENZEL, E.W.: Early experience and the social development of rhesus monkeys and chimpanzees; in NEWTON and LEVINE Early experience and behavior (Thomas, Springfield 1968).

RIESEN, A.H.: Nissen's observations on the development of sexual behavior in captive-born, nursery-reared chimpanzees; in BOURNE The chimpanzee, vol. 4, pp. 1–18 (Karger, Basel 1970).

ROGERS, C.M. and DAVENPORT, R.K.: Chimpanzee maternal behavior; in BOURNE The chimpanzee, vol. 3, pp. 361–368 (Karger, Basel 1970).

ROGERS, C.M. and DAVENPORT, R.K.: Effects of restricted rearing on sexual behavior of chimpanzees. Develop. Psych. *1:* 200–204 (1969).

Author's address: Dr. CHARLES M. ROGERS, Department of Psychology, Auburn University, *Auburn, AL 36830* (USA)

Symp. IVth Int. Congr. Primat., vol. 1: Precultural Primate Behavior,
pp. 192–225 (Karger, Basel 1973)

Leadership and
Communication in Young Chimpanzees[1]

E.W. Menzel, jr.

Department of Psychology, State University of New York at Stony Brook

All social life depends on the ability of individuals to coordinate and
regulate their actions with respect to each other and to the environment. A
process by which this coordination is achieved is communication. Of necessity
communication processes involve some sorts of signals – vocalizations,
gestures, scent emissions, and so on. In man and to some extent in other
primates [see Hewes, Itani, van Lawick-Goodall and Stephenson, this
volume], considerable learning on the part of individuals is required before
these signals become ritualized and used in the proper contexts as a formal
social code. Thus, some communication systems – especially, of course, verbal
language – are not only processes for bringing about cultural behaviors and
transmitting them to others, but are also forms of cultural behavior in their
own right. In this sense, Hall [1959] is correct in maintaining that 'Culture is
communication and communication is culture'.

The topic of communication may be approached through a study of
signals as such or through the larger facts of sociological coordination, for
instance, group travel. In theory these two approaches are complementary
and should lead to the same conclusions in the final analysis. But in practice I
agree with Carpenter [1964], Kummer [1971], and Crook [1970] that the
detailed examination of the ethological 'elements' of communication has been
overemphasized and that if we are interested in the sociological 'gestalts', we
must find methods that will deal with them more directly.

There is of course no substitute for a signal-oriented approach if one's

1   Data collection was supported by NIH grant FR-00164 to the Delta Regional Primate
Research Center, and writing by NSF grants BO-38791 and GU-3850 to the Psycho-
biology Program of the State University of New York at Stony Brook. I thank Palmer
Midgett, jr. for his assistance with the experiments.

interest is in signalling behavior *per se*. Thus, a majority of primatological studies focus attention on selected classes or subclasses of potential signals and examine how they serve to coordinate and regulate behavior in a given society. According to MARLER [1965], the three basic questions which one might ask here are: What is a given society's repertoire of visual, auditory-vocal, tactile, olfactory (etc.) signals? Under what circumstances does each class or subclass of signal occur? And, how do the other members of the society respond to these signals? In general, naturalists are most interested in what coordinative functions each class of signal typically *does* serve in a wild population [e.g., ALTMANN, 1962; MARLER, 1965, 1968; VAN LAWICK-GOODALL, 1968]. Experimentalists are also interested in what functions vocalization, or gesture, or visual signals *could* conceivably come to serve under optimal conditions of training – that is, in the upper limit of communicative capacity in a given species [HAYES and HAYES, 1954; GARDNER and GARDNER, 1971; PREMACK, 1971; HEWES' review, this syposium].

Sociologically-oriented approaches to communication have not as yet been very well articulated in primatology, but their distinguishing feature is that they focus attention on an obvious case of coordinated group action in a given society and then work backward to determine whatever signalling systems and other variables bring the action about. Discussions of leadership, cooperation, and group organization [e.g., CARPENTER, 1964; CROOK, 1970; KUMMER, 1968; SCHALLER, 1964] often proceed in this fashion. The model studies, however, are no doubt VON FRISCH'S [1955] experiments on the foraging of bees.

For many years beekeepers and hunters had claimed that if one bee finds a new source of food, other members of the same hive (but not members of different hives) soon start to arrive on the scene. How would such apparent group coordination be possible without some form of communication, in the broadest definition of that term [ALTMANN, 1967; WIENER, 1961]? VON FRISCH'S [1955] initial problem was *whether* bees are informing each other of the food, whether, that is, some type of behavioral cue from other bees is necessary, and common response to some 'external' variable such as odor of food is not sufficient, to account for the group response. The answer naturally gave rise to more specific questions:

Who is controlling the group or furnishing its members the necessary information? This was answered by determining which bees had perceived the food themselves and which bees came into contact with those who had perceived the food and then by examining the group response.

What information about the environment does the original forager retain, and what does she convey to others? This was answered by varying the nature and location of the goal and then examining the consequent variations in the behavior of those who perceived the goal for themselves and those who were informed only by this individual. To the extent that those who have not perceived the food for themselves can get there as well on their first trip as their informer can on her second trip and that they vary their behavior according to the nature of the food supply as well, it would appear that signals from an informed animal are informationally equivalent to 'having seen for oneself', the best criterion we have of a sign process [cf. MENZEL, 1969; MOWRER, 1960; PAVLOV, 1927].

How are the bees communicating what they are communicating? Here for the first time it becomes essential to examine precisely the ethological details of the behavior of the 'informer' and the 'informed', when they come into contact with each other, and to determine which aspects of the informer's behavior the other bees use as the social cue.

It might be mentioned that VON FRISCH knew what the bees could accomplish by way of communication (and from this knowledge he surmised that they must have some sort of means of communicating about the environment) long before he discovered the function of their dances. In principle he might have discovered the sociological facts of foraging from a signal-oriented examination of dance behavior *per se*, but as a matter of biographical record he seemed to proceed in the opposite fashion.

There follows a brief account of my own experiments on chimpanzee communication, which used procedures and a logic analogous to those of VON FRISCH. Portions of this work are more fully described and quantified elsewhere [MENZEL, 1971b, in press].

The Fact of Coordination, and the General Problem

No one doubts that most species of primates live and move in cohesive and stable social groups, constantly adjusting their movements with respect to each other and the environment. In the course of their travels, the group-as-a-whole locates food, water, shelter, and all other basic necessities and manages to avoid dangerous situations and predators. Although the members of a group sometimes act as if they were in conflict about which direction to take next, these conflicts are mild and infrequent; and usually all animals set off together in the same direction [KUMMER, 1968, 1971]. For wild chimpanzee groups, a high degree of cohesiveness is seen clearly only in mother-offspring units; but there is increasing evidence that earlier reports [GOODALL, 1965; REYNOLDS and REYNOLDS, 1965] underestimated the stability and coordination that exists in larger populations [VAN LAWICK-GOODALL, 1971; ITANI and SUZUKI, 1967].

The fact of cohesion and coordinated travel can easily be observed in captive chimpanzees, provided only that their cage is large enough and the animals are young enough and have lived together long enough to form attachments to each other. Indeed, the cohesiveness of the group of juveniles that I observed in a 0.37-hectare (almost one acre) enclosure at the Delta Regional Primate Research Center in Louisiana was so striking and their ability to decide instantly among themselves which way to go next was so impressive that I almost began to believe that they possessed a 'group mind'.

At first I was most interested in *why* these chimpanzees ordinarily travelled and foraged as a spatially cohesive band rather than in some other fashion (e. g., like bees). I analyzed this problem in part by examining their early rearing experiences, their past associations with particular companions, and the immediate stimulus conditions under which the group would disperse about the enclosure or come even closer together. As it turned out, the best predictor of how closely any two animals would follow each other was how long they had lived together [cf. 'family groups' in the wild, VAN LAWICK-GOODALL, 1971]; and the only stimulus variable that could cause the group to scatter across the whole enclosure (30.5 × 122 m) was either food or some object that served as a sign of food [cf. REYNOLDS and REYNOLDS, 1965]. However, these problems are of secondary interest to the student of communication; and since the details of the studies are presented elsewhere [MENZEL, in press], I shall not deal with them here. The more important problem is *how* the animals were coordinating their movements.

I could see relatively few instances of any clear signalling behaviors of the sort reported in studies of leadership [SCHALLER, 1964; KUMMER, 1968], co-operation [CRAWFORD, 1936], 'language' training [GARDNER and GARDNER, 1971], or emotion-provoking situations [VAN LAWICK-GOODALL, 1968]; there was perhaps only one good vocalization or hand signal, etc., detectable for every ten cases of obviously coordinated travel. No single individual seemed to serve as the leader, either. Nevertheless, short of extrasensory perception, there seemed to be only two plausible alternatives. Either the animals had an uncanny knack for perceiving the same toy or piece of food or snake in the grass at the same time, or they were getting their cues from each other; and any individual was capable of directing the group to things that he alone had perceived and of getting them to respond very much as they would have if they had seen the toy or food or snake for themselves. Observation alone was inadequate to permit a decision between these alternatives.

Are the Animals Communicating about the Environment?

Some investigators [e. g., ALTMANN, 1967] use the term 'social communication' in such a broad way that it applies to almost any sequential dependency or correlation between the actions of two animals; others [LYONS, 1972] use it in such a restrictive way that, almost by definition, it would be difficult if not impossible to demonstrate 'real communication' in a nonverbal animal. MACKAY [1972] has very carefully specified the criteria that must be considered in arriving at any useful and nontrivial definition. His analysis suggests that it is less important to be able to apply the magic term to one's data than it is to examine the degree of correlation between animals, the precise conditions under which the correlation breaks down, and the causal mechanisms underlying the correlation.

Table I shows the degree to which the spatial locations of each animal in an aggregation of eight were correlated with the spatial locations of each other animal, under routine conditions. It can be seen that most of the correlations are very high and positive in sign. Obviously something is going on between the animals. When A moves, B does too, or vice versa; and an observer could easily predict one animal's movements from those of almost any other. (On the broadest definition of the term communication, we have already answered the question of *whether* the animals are communicating.) The longer a given pair of animals had lived together before the time of the test, the more closely interrelated were their movements. As the animals grew older and more independent, these effects decreased markedly; but even at adolescence (several years after the data of table I were collected), the effects were still discernible.

Equally important, other tests also showed that an observer could predict the movements of the whole group from a knowledge of what potential goal objects were in the field. In one series of tests [MENZEL, 1971a, in press] the animals were locked into a small release cage on the periphery of the one-acre field, from which they could not see into the field. An experimenter then entered the field and placed a new toy at some randomly designated location, positioning it in such a way that it would be clearly visible from the door of the release-cage. Then the group was turned loose. Usually the animals travelled closely together. Over any series of ten or more tests, the correlation between the direction of travel of the group-as-a-whole and the direction of the novel

Table I. Baseline data showing the extent to which given pairs of animals move together: spatial association as measured by the linear correlation coefficient (Pearson r) between two animals' locations over 1,500 time intervals (30 per session for 50 sessions)[1]. Animals were designated *a priori* as belonging to a given 'subgroup' purely on the basis of how long they had lived together before the test [see MENZEL, in press, for details]

Sub-group	Chimp	I			II			III	
		Shadow	Bandit	Belle	Polly	Bido	Libi	Rock	Gigi
I	Shadow								
	Bandit	0.82							
	Belle	0.94	0.84						
II	Polly	0.80	0.86	0.80					
	Bido	0.65	0.70	0.68	0.77				
	Libi	0.64	0.70	0.67	0.77	0.98			
III	Rock	0.46	0.47	0.47	0.51	0.43	0.42		
	Gigi	0.47	0.47	0.48	0.53	0.43	0.43	0.94	

1 Statistical note: with two-dimensional (X, Y) locational data and T time intervals, the relationship between the average squared distance between two animals (i, j) and their locational Pearson r is

$$\frac{1}{T} \sum d_{ij}^2 = \sigma_i^2 + \sigma_j^2 + (\overline{X}_i - \overline{X}_j)^2 + (\overline{Y}_i - \overline{Y}_j)^2 - 2 r \sigma_i \sigma_j,$$

where $\sum d_{ij}^2$ is the sum of the squared distances between two individuals i and j over all T intervals and σ^2 is the variance of an animal's locations around his own mean location $(\overline{X}, \overline{Y})$ or $1/T^2$ times the sum of the squared distances between all possible pairs of his momentary location-points.

object from the release cage was about r = 0.95; but the magnitude of the correlation naturally declined if the chimpanzees were given repeated encounters with the same objects and lost interest in them.

These correlation coefficients are admirably quantitative; they can be transformed into still more scientific-sounding measures of information transmission [ATTNEAVE, 1959]; and they furnish an objective, parsimonious, and systematic way of summarizing a large amount of data. I prefer them to a purely verbal qualitative form of description. Nevertheless, the student of animal behavior should not be fooled into thinking that they tell us much more than we knew before. The correlation coefficients simply restate in numerical form, and more precisely, the original 'fact' of group cohesion and coordinated travel with respect to objects that we have already described in intuitive terms.

They do, however, make the important point that there is more 'communication' and 'transmission of information' (in the broad sense) going on in the interactions of a group of animals than is likely to be suggested by most ethological catalogues of the molecular aspects of individual behavior. Unless a description of the molecular details of communication can fully account for the total amount of information transmitted at *all* levels of analysis (instead of simply taking much of this information for granted), it is neither a complete nor an adequate description of group communication processes.

How can we now become more analytical and determine what proportion of the overall correlation between the group travel angle and the direction of their goal is attributable to A's control over B, to B's control over A, to the joint action of a third variable on both A and B, and so on? This task is more difficult to accomplish with chimpanzees (even captive ones) than with bees, unless one is willing to resort to some experimental interference with their normal foraging and travel patterns. It requires that we know when A perceives the goal independently of B, when B perceives it independently of A, and so on. With bees, which forage at long distances from each other and then return to a central hive, this knowledge can be gained with minimal interference. All VON FRISCH had to do was to put out some food and then place an identifying mark on those individuals who came to the food. With my chimpanzees, who were seldom far apart from each other and who had no home base they returned to regularly, it was seldom possible to tell through observation alone who got the information independently of all others. Therefore, I resorted to using a slightly modified version of the delayed response experiment.

The sole differences between these delayed response tests and our previous tests with novel objects were two. (a) The goal object (that is, the object we introduced into the field while the chimpanzees were locked into the release-cage) was carefully hidden, so that it would be invisible when the chimpanzees were turned loose and extremely difficult to find by chance. (b) Only certain individuals (here operationally-designated as potential 'leaders' on a given trial) were given any cue as to the nature and whereabouts of the hidden object. We gave selected individuals a cue by carrying them over to the object and permitting them a direct look at it or by carrying them a shorter distance and giving them a social signal (such as pointing). Then we returned them to the group; the other animals ('followers') had to get the messages from these leaders. Altogether we used eight chimpanzees in the experiments; but only five could be tested as leaders, because the others became so upset at being separated from their companions that they refused to let us carry them off. A description of our three types of tests and their basic results follows.

First, all five chimpanzees who were tested as leaders were capable of remembering the exact location of the object that we had shown them and of inducing the rest of the group to go along with them to this object. Since effective performance was noted even in the very first trials of each leader's tests, it would appear that no special training on our part was necessary; presumably the animals had already trained each other.

Second, performance was clearly a cooperative venture, in which the leader was as dependent on the followers for getting to the goal as they were dependent on him for knowing where to go. In some tests we showed the food to one leader; and then, instead of returning him to the group and releasing everyone, we placed him in a separate release cage and turned him loose alone. Given this chance to get the goal for himself, the leader whimpered, screamed, defecated, and begged at the observers with an extended hand or tried to open the cage to release his companions. (There were, however, definite changes in this behavior with age and experience. Over a period of several years all animals showed a decreased tendency to be upset at being left alone and an increased tendency to get the goal for themselves if it was not too far away from the release cage.) As soon as the followers were released, the 'leader' ran to them, embraced a preferred companion, and then set off for the goal. In no experiment did these juveniles ever set off without an occasional glance backward at their companions; and even a fast running start would stop cold unless there was a good following.

Third, control tests showed that neither the leader's nor the followers' ability to get to the hidden goal could be explained by their response to cues available after the group was released (odors from the goal object, inadvertent behavior on the part of the observer, signs of our trail in the field, etc.). If we hid the goal object but showed it to no one, the chimpanzees either did not search the field at all; or they searched 'at random', in which case they were lucky to find one goal in 30 trials even after covering the whole field.

Taken together, these three types of tests (leader with followers, leader alone, and no informed leader) clearly show that any single informed animal, A, can control the travel of the rest of the group $(B + C... + N)$; that group factors in turn control A's behavior; and that the correlation between the group majority's travel route and a distant goal cannot be explained by the common response of all animals to some external (nonsocial) variable. In other words, information-exchange of some sort between animals is necessary. This more than satisfies a moderately stringent definition of 'communication'. To satisfy a most stringent definition [MACKAY, 1972] we would, however, have to break down the observed correlations still further and show that the leader's behavior is goal-directed and that the followers perceive that the leader's behavior is goal-directed. This will be done in later sections of this paper, especially in the one on communication of direction. Anticipating the data, I believe that there is convincing evidence that the leader varied his behavior in almost every possible fashion that would simultaneously get him to the hidden objects and prevent him from having to go there alone (including, sometimes, tapping other animals on the shoulder, tugging at them, 'presenting' the back, or even biting a companion on the neck and dragging him along the ground by a leg in the direction of the hidden object; see MENZEL, 1971b). And the followers in turn varied their behavior in such a way as to indicate that they did not merely follow the leader for his own sake, but also used his behavior as a cue to the distant hidden object. By any objective criteria of goal direction and purposiveness [e.g., MACKAY, 1972; TOLMAN, 1938] both the leader's and the followers' behavior can be so described.

In the same spirit, one might ask further, whether, in effect, the leader knew that the followers knew that his behavior was a sign of the presence and direction of the goal and varied his actions accordingly. MEAD [cited by MORRIS, 1955] suggested that this complex sort of relationship is the crucial criterion that a language-like process is at work. In animals this relationship is perhaps best assessed by asking whether the leader ever appeared to vary his behavior in such a way as to *misinform* the others. Some linguists [e.g., J.C. MARSHALL, cited by LYONS, 1972] have indeed suggested that 'intention to misinform' would be the best single piece of evidence for communication in the very strict sense and that 'prevarication' is one of the 'design features of language' [HOCKETT, 1960; ALTMANN, 1967].

In the present tests [see also VAN LAWICK-GOODALL, 1971] there were occasional attempts to withhold information, if not to actively mislead. Consider our chimpanzees, Belle and Rock. These two animals were relative strangers to each other, and Rock was highly dominant over Belle and all other members of the group. When they were used as the leaders in the paired comparison tests, where we showed each of two leaders a different pile of food on a given trial and they and the followers had to choose which way to go, Rock nearly always attracted the smaller followings and Belle the larger. In fact, Rock often had no following at all. The result was that he (who, like all the others, seldom went far without at least one companion) had either to wait for the followers to return from their trek with the preferred leader or to go with the mob in the opposite direction from 'his' goal and then lead them all the way across the field after they were finished with the first goal. After several days, Rock started to follow Belle and chase her from her food when she uncovered it. This behavior understandably caused Belle to go more and more slowly, and finally she stopped uncovering the goal and sat on it instead until Rock left. Some days later, however, Rock would no longer leave. He shoved Belle aside whenever she sat down and searched the place in which she had been sitting. Or, in other cases, he wandered off, not looking back at all until he was at least five meters away; and then he usually caught Belle in the act of uncovering the goal object. Over a period of a month or two, things kept progressing in this fashion until Belle would sit as far as 10 meters away from 'her' goal so long as Rock was in a position to see her; and Rock in turn searched up and down the general angle of Belle's line of bodily orientation, intensifying his search and adjusting his location whenever she showed any increase in 'nervous' movements. Eventually he found goals that were 10 meters or more from Belle. Unfortunately, we had to terminate the test because Rock became ill and had to be hospitalized. At the end of the test Belle was still not a good enough liar to keep from giving the game away by fiddling with an object, scratching, whimpering, shifting about restlessly, or standing up when Rock got very close to the hidden goal. However, on several occasions she actually started off a trial by going in the opposite direction from 'her' goal and then (when Rock started his search) doubling back rapidly. None of these behaviors on the part of Belle were seen on trials in which Rock was kept out of the test, and Rock seldom if ever poached on the other leaders or followed Belle if she had not been shown a goal.

The notion that animals express all their emotions overtly and cannot withhold information from each other [HEBB and THOMPSON, 1954] is thus clearly not true [see also LORENZ, 1970; MENZEL, 1966; VAN LAWICK-GOODALL, 1971]. For chimpanzees, the learning of 'negative information' or of an inhibitory response might, however, be much more difficult than the learning of 'positive information' or approach responses [DAVENPORT et al., 1969; GARDNER and GARDNER, 1971; MENZEL and DAVENPORT, 1962; THOMPSON, 1954]; and in spatial problems the difficulty of learning increases as a function of the degree of deviation required from a direct and immediate approach to the goal [e.g., KÖHLER, 1925].

Who Controls the Group's Travel?

As I have already indicated, any of the five individuals in the group of eight whom we were able to test formally as a potential leader (including Rock and Gigi, who were introduced as relative strangers) was capable of inducing the others to follow him to a goal that only he had seen. From their behavior under ordinary circumstances, I am certain that the three remaining animals (Libi, Shadow, and Polly) were as capable of 'leadership' as those we tested formally. Thus, to a considerable degree, the members of the group were interchangeable as leaders or followers, as senders or receivers of travel directions; such 'interchangeability' is of course another distinguishing feature of communication [HOCKETT, 1960].

This is not to say that these chimpanzees were as homogeneous or as completely interchangeable as forager bees. In tests where only one leader was shown a goal, there were clear differences in the relative speed with which the group moved out. These differences were partly caused by nonsocial factors, such as the leader's running ability and apparent motivation for a particular class of object, and partly by social factors, such as the leader's dependence on having a close following and the followers' tendency to straggle more slowly behind a non-preferred leader (in which case any leader would slow down) or keep close to or even ahead of a preferred leader (in which case any leader would speed up). By varying the presence or absence of followers, the number of followers, and the identity of the followers in different tests, we could cause each leader's rate of travel to vary from no travel at all through a very slow and cautious move to a trot or run (which usually produced a 'pack-run' in the group); nevertheless, certain individuals were characteristically faster than others.

In tests where each of two leaders was shown a different goal and the animals were forced to make a choice of travel direction, some pairs of leaders (e. g., Belle and Gigi) nearly always split from each other and went in opposite directions at the same time; other pairs (e. g., Belle and Bandit) more often pulled each other this way and that and wound up going together, first to one goal and then the next. There was a linear hierarchy in 'leadership ability', as measured by the number of trials on which the group majority went to a given leader's goal, the average number of followers accompanying each leader, and the average difference between the travel angle of the group-as-a-whole and the direction of each leader's goal. The existence of leadership hierarchies was, incidentally, questioned in earlier field studies of chimpanzees; but the more recent and thorough examination of this problem on the part of the Kyoto University investigators has produced convincing evidence for leadership [ITANI, personal communication]. No single individual invariably serves as 'the leader', as in wild gorilla groups [SCHALLER, 1964]; rather, for natural chimpanzee groups as well as for my laboratory group of juveniles and for human beings, it would be more accurate to talk of leadership *behaviors* or *roles* that all individuals other than very young infants display on some occasions [GIBBS, 1969; see also KUMMER, this symposium]. To some degree, most individuals can act as leader if the momentary occasion demands; there is at least partial interchangeability of individuals.

There is, of course, no necessary relation between 'who controls the group' and 'who communicates to whom'. Control, of necessity, implies some sort of communication (in the general sense), but animals can certainly get information from each other without being able to induce each other to follow. Insofar as one can generalize from my very small sample of individuals, it would appear that in order to be a 'good leader' (i.e., in order to attract a large number of followers to a hidden goal), a given individual had to possess all of the following qualities simultaneously. (a) He must be on moderately good terms with the followers. (b) He must be independent enough initially to move out on his own.(Even the most 'independent' of my animals seldom moved out without occasional glances backward to make sure they were followed by at least one companion.) (c) He must be intelligent enough to know the lay of the land and to remember the location of previously encountered food supplies, etc. (d) He must share the rewards, to some extent at least. Especially in food-getting tests, there was a substantial correlation between leadership preferences and the frequency with which food-sharing occurred under ordinary circumstances.

Qualities (a) and (d) depend heavily upon how long a potential leader had lived in the same cage with his followers (i.e., familiarity). Quality (b), ability to move out on one's own, is determined above all by the age and size of the animal [see also MENZEL, 1969]. All of my animals possessed quality (c), but presumably very young infants would not. I found no evidence for the tacit assumption of many researchers that leadership is the prerogative of the most dominant male. Indeed, the least preferred leader was Rock, the dominant male; and the most preferred leader, Belle, was one of the least dominant females.

Figure 1 shows the relationship between the data of the formal 'leadership' tests and the data from table I, from which we started our discussion of communication. It can be seen that 'who follows whom' in a formal test where one individual is shown a hidden goal is predictable from 'who associates with whom' under ordinary circumstances. It might be noted that the 'hidden goal' in these particular tests was food and that leadership hierarchies differed somewhat according to the class of object. For example, Rock was the least preferred leader when food was involved (his relative unfamiliarity, his high dominance status, the fact that he seldom shared food, and the fact that the

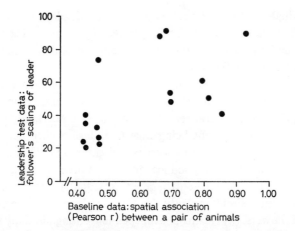

Fig. 1. Can the preference of a follower for a given leader in a foraging test be predicted from the spatial association of the pair under ordinary circumstances? The 16 points on the graph show the preferences of Shadow, Polly, Libi, and Bido (who were always tested as followers in this test) for each of the four leaders in this test (Belle, Bandit, Gigi and Rock). The abscissa is based on the data of table I; the ordinate is based on the percentage of trials in which one leader (rather than another) was followed.

foods we presented were not widely dispersed, as they would be under natural conditions in the wild, were all undoubtedly of importance here). But Rock was highly capable of leading the group to 'mob' a snake or a stranger.

What Information Does the Leader Retain and Convey to Others?

One of the most elementary but basic problems of sign learning is how an object (e. g., a banana) can affect the behavior of an animal almost as much when it is 'out there' at some remote point in the environment as when the animal is in direct physical contact with it [PAVLOV, 1927]. In the same sense, the problem of memory arises when the effectiveness of the object still persists even after the animal can no longer perceive it. Still further, our present problem of 'social communication about the environment' arises because an object which even the leader can no longer perceive nevertheless affects the responses of a whole group of animals.

In other words, communication about the environment is a special case of sign learning in which information must, in effect, cross the ether from the goal object to one animal (the leader); be processed, retained, and acted on by him; and then cross the ether again to other animals (the followers) and be processed and acted on by them. If we knew precisely what information gets from each point of the ecosystem to the next, how it gets there, and how it is transformed and modified or attenuated at each stage of this process, we would have solved most of the problems of chimpanzee psychology.

The question now before us is what the leader knew about the hidden goal object from the cue that we had previously given him and (more important) what the followers in turn learned from the leader. To answer this, I examined how controlled variations in the goal objects or in the cues that we provided the leader affected the response of the leader and his followers. To the extent that the animals can respond differentially to goals that vary in only one attribute, I assume that they must possess some information regarding this attribute.

a) The presence and absence of an object. The leaders usually travelled in a direct line to places in which they had seen a novel or salient object, and they either did nothing or followed someone else when they had been shown an 'empty' pile or nothing at all; thus they obviously retained information regarding the presence vs. absence of an object.

The presence of an object can also be said to have been communicated to others. Not only did the followers stick close to a leader who was headed for hidden food, but to do so, they also often abandoned a smaller food reward that was directly visible at the time of response (and that was readily approached on control tests). In fact, on 88% of the trials in one test, a leader going to four pieces of hidden food was accompanied by more animals than went to a visible single piece of food. This result cannot be explained merely by the gregarious tendency or social dependence of chimpanzees, for the group readily split if the rewards favored such a response or if two leaders pulled vigorously enough in opposite directions.

The *absence* of a special incentive on a given trial was, however, much more difficult to convey to a chimpanzee. Followers showed no evidence of being able to discriminate between trials on which the leader had been shown no goal vs. trials on which he had been shown a goal but (for any reason) failed to respond to it himself. (This is not to say that followers could not act on less than the full response of 'going all the way to the object and uncovering it' or that they did not occasionally pull a leader to his feet when he dawdled.)

In one experiment on the communication of direction we used ourselves as 'leaders' and gave the chimpanzees one social signal (walking a few steps, stopping, and orienting and gesticulating vigorously in one direction) as a cue of food and another social signal (walking a few paces in a given direction but then stopping and turning in a circle and orienting all over, as if we had seen nothing) as a cue of no food. Following this signalling, we removed ourselves from the situation. If both cues were presented on a given trial, the chimpanzees went in the 'correct' direction at better than chance frequencies. That is, they discriminated which of the two goals was relatively *better*. But when either the 'food cue' or the 'no food cue' was presented by itself, their responses did not differ either in the frequency of travelling or in the accuracy with which they selected the same direction we had 'pointed out' by our walking in a given direction.

b) Direction. That the leaders remembered the exact location of a hidden object that they had previously seen is already obvious. Indeed, they could proceed to any location from any starting place in the field. In many cases one leader accompanied another to his goal first and only after this went to his own goal; nevertheless, he went to it almost unerringly. In one test, we showed a leader 18 different piles of food on a single trial, and varied these locations randomly from one trial to the next. The best of the leaders (Belle) sometimes found all 18 piles. No leader ever followed the trail we had used in showing the foods. This observation, together with the fact that we had carried the animals to show them the food and the further fact that followers found food only by searching alongside of the leader, rules out the possibility that performance was based on cues such as odor, motor set, inadvertent behaviors on the part of the observers, etc. The term 'cognitive map' [TOLMAN, 1938] seems highly appropriate.

Altogether, I conducted seven experiments to test whether or not the chimpanzees could get directional information from another animal and how great a displacement between cue and reward was possible before efficient performance would break down. All seven experiments yielded positive results. To illustrate, I consider here the first experiment and one of the last.

(1) Food is hidden in one of four locations about 50 meters from the release cage and shown to Bandit. Then Bandit is returned to the group, and several minutes later everyone is turned loose together. Over a series of several trials the food is randomly varied from one of these four locations to the next; but since the same locations are used from day to day, the followers are given a chance to learn that when Bandit heads in a given direction, he is headed for the goal in that direction. Will they learn to anticipate his destination, and beat him to it? And at what point in the travel sequence do they seem to anticipate the destination?

On the first day or two Bandit, not surprisingly, was the first animal to reach the hidden food on almost every trial. By the third day, however, the followers started to jump ahead of him in the last few meters. On subsequent days they broke from him at progressively more remote locations, until finally they ran out of the release cage first on a given trial and waited for Bandit to emerge; and then, as soon as he merely looked one way or the other, they were off like a shot. If Bandit got any food at all after the fourth or fifth day, it was usually obtained by begging from his 'followers'. He understandably started to stall, and the followers either set off on their own in all directions or tried to drag Bandit out of the release cage.

Changing the leadership from Bandit to another animal disrupted the followers' success for a few trials (until they learned to watch the new leader instead), and changing the locations disrupted their success for two or three days (until they learned to stop 'jumping the gun' and racing for an old location as soon as the leader oriented in that direction).

(2) Food is placed at random at any point in the field, but left uncovered so that it can be spotted from perhaps five meters. No one is shown the food; instead, one experimenter holds the 'leader' at the door of the release cage while another gives a social signal, such as pointing in the direction of the food. Then the experimenters leave the field, and the animals are turned loose only after a delay of several minutes.

Results: The animal to whom we had given the signal on a given trial searched up and down the field in the appropriate direction. Sometimes no animal could find the food; and the group wandered off, only to return to the same region to resume the search. In contrast, on control trials the animals searched at random with respect to the hidden goal.

Not only could the leader learn the direction from us in a single exposure and remember our signal for several minutes; but his followers could take the cue from him, search ahead of him, and (as often as not) get to it first. One follower, Bido, even developed the strategy of racing ahead of the leader, climbing up the nearest tree that lay in that direction, and scanning the field from a height of 10 meters or more. Her success in getting the food was quite good until others started to watch her if she came down very rapidly.

In these tests, where no animal had precise information as to the exact location of the object and where (because the food lay uncovered) general searching activity on the part of the followers was reinforced, we saw many instances in which the whole pack unintentionally led each other in a wrong direction. If one animal for any reason started to move in a slightly different direction and more rapidly than the others, someone would watch him closely and try to head him off; the instigator of the move would then in turn seem to take this as a cue that the other animal knew where to go and would try to keep ahead of him; then still others would see these two animals and start after them; and so on, in a 'vicious circle', until the whole pack of followers was racing down the field and checking every likely-looking hiding place along this path. As soon as someone stopped, however, the animal who had been attending to him would also stop; and his follower also would stop; and so on, until all animals stood around eyeing each other. They then returned to the place where the original search had left off. The leader (to whom we had given the cue of direction) usually joined in this activity only if he had already searched to the limits of the field in a given direction without finding the food or if our cue was a relatively ineffective one.

In general, then, I believe that signals which carry information regarding the probable direction of a distant goal can be passed along a chain of at least three chimpanzees (if not a whole group), and it is not necessary that *any* animal has seen the goal for himself. To be anthropomorphic, chimpanzees can bear rumors.

c) Distance and location. It is a common observation in many types of experiments that animals will pull harder, run faster, respond more frequently, and show a preference in a choice situation for a near versus a far goal object. These effects are enhanced if the object is visible at the time of response, but they are commonplace even for hidden objects.

Our chimpanzees were no exception to the rule. If a hidden pile of food was very close (say, less than 10 meters) from the release cage, the leader seldom tried to get the rest of the group to come along; he struck off by himself. If the goal was somewhat farther off, the leader moved more slowly at first and tugged vigorously at the followers; then, as he got closer, he broke from his companions and went into a goal spurt. If the goal was at the extreme end of the cage, the leader rarely went alone; and unless others followed readily, he might tug at them once or twice and then simply quit. Most of the leaders' failures to get started at all occurred with goals that were either distant or that required the animals to pass through a non-preferred area of the field. In the test where we showed a leader 18 piles of food, he generally went to one of the nearest piles first and showed a clear preference for a 'least distance' route overall.

When objects other than food were involved, some of these distance effects were reversed. Thus, for example, a leader slowed up and stayed even closer to a companion (instead of speeding up and leaving others behind) as he approached a place where a snake lay hidden in the grass.

Such variations in the leader's behavior were certainly sufficient to provide a human observer with a fairly good cue as to the distance between the leader and the goal object, especially in cases where the leader oriented downward from a tree toward the ground or where he accelerated suddenly and headed directly toward a highly prominent object, such as a tree in the middle of an open area. A further social basis for predicting the exact location (i.e., both distance and direction) of an object would be seen when several animals oriented simultaneously toward the same point from different positions in space, thus allowing the observer to 'triangulate' from the several directional cues onto this point [cf. ITANI, 1963].

From general observations I see no reason to believe that follower chimpanzees are any poorer than human observers at predicting, from the leader's behavior, the location of a goal object. If a leader accelerated suddenly toward a likely-looking hiding place, others dove for the same place; and if the leader slowed down gradually and showed piloerection while staring straight at a clump of grass, they, too, stared at approximately the same place and appeared cautious toward that place. This is not to say that the leader had in effect one specialized set of 'words' which conveyed distance information *alone*. Possibly a follower could not tell for sure from, e.g., the leader's intensity of tugging or his speed of running whether the object was a highly preferred goal that was far away, a moderately preferred goal that was fairly close, or a mediocre goal for which the leader happened to be highly motivated because he hadn't eaten his dinner last night. My data on this point are less than definitive.

   d) Class of object. As the discussion above suggests, both the leader and his followers discriminated between trials in which a desired object such as food was presented and trials in which the hidden goal was fear-inducing. In some tests we showed the leader either a pile of food or a dead snake (or stuffed alligator, etc.) and then completely removed it before releasing the group. Only one such critical trial was presented in a given day. On other (baseline) trials we used food as the goal object and left it where the leader had seen it.

   On the 'food removed' trials, the typical reaction of the group was to run at their normal speed and then manually search through the grass all around the place in which the leader had seen the food. On the 'snake removed' trials, not only the leader but also the followers often emerged from the release cage showing piloerection and staring in the direction of the snake that was no longer there. They moved slowly, acting more and more cautious as they drew closer to the exact place in which the snake had been seen. Then one or more chimpanzees – including followers, who had never seen the snake – hooted and threw sticks or grass or slapped at the ground. Often a follower was the first to attempt to hit the mark. Clearly, both the leader and the followers expected something good and something bad, respectively, on these two types of tests [cf. TINKLEPAUGH, 1928]. Furthermore, in both types of tests the followers did not merely match their emotional state to that of the leader; they directed an appropriately varied response to the *place* 'out there' that had contained the object of the leader's excitement.

   The theory that there is some sort of fundamental distinction between propositional communication (communication about objects) and emotional communication [e.g., KÖHLER, 1925; CASSIRER, 1944; MARLER, 1965; PRE-MACK, 1972] thus misfires. In our data, the form or quality of the animals' molecular reactions can obviously be termed 'emotional' if one so chooses; but how would the followers know *where* to orient or toward what *environmental features* to direct their reactions, without 'propositional', object-related information? To explain the behavior of these chimpanzees on the basis of 'purely emotional communication' would be to leave them suspended short of the goal or emoting all over the place. Emotional communication and propositional communication are not mutually exclusive categories of behavior.

Indeed, I rather doubt that it would be possible for one chimpanzee to get another to imitate his cautious responses toward a hidden snake if there were no objects (such as a clump of grass or a hole in the ground in which a snake might hide) to serve as contextual supports for such behavior. To illustrate the importance of 'contextual supports' we conducted another experiment. On some trials we pointed a rifle at a bush or a particular spot in the tall grass (which could easily contain a hidden snake) and acted as if we were orienting toward something fearsome. The chimpanzees usually approached that spot with caution, stared from a bipedal posture, and threw a stick at the same spot. On other trials we gave the animals the same sort of cue, except that we pointed the rifle all around in the air (at no place in particular) or aimed it at a place on an open patch of ground (which could be visually surveyed at a glance and held no hiding places). Here the chimpanzees showed no emotional reaction of any sort. They looked briefly at us, and at the place we aimed, and then went back to whatever they were doing.

e) Relative quantities of food. Chimpanzees run faster, work harder, and are more capable of inducing a following for a preferred object than a non-preferred one; this pattern provides an obvious mechanism whereby several potential leaders who get together after having seen different food supplies might be able to 'tell' each other and the group which of these several supplies is likely to be the best. The principle is simply: 'Go with the most eager leader. If others are more eager than you, there is good reason for it; so forget your own goal or leave it until later.' The eagerness principle of course operates only in a society in which animals can discount the general status of a given leader and choose between leaders according to what they *do* rather than (only) who they are. And the eagerness principle is a sufficient basis for always getting to the best available goal only if all animals either share the same value systems and display the same level of motivation for a given object or can learn to take into account each other's idiosynsrasies.

PREMACK [1972] argues that without a symbolic language system, animals must share identical preference systems and levels of motivation before they can inform each other about objects and events. I doubt this. A mother chimpanzee, for example, usually knows from her infant's behavior who or what frightened it, even though she does not fear the same thing or imitate the infant's reaction. In my tests, followers seemed in some instances to be more motivated for Bandit's goal than Bandit was himself, an event which I would not expect from PREMACK's hypothesis. If Bandit (usually after eating plenty of food) failed to lead out after having been shown a hidden goal, the 'followers' sometimes dragged him out of the release cage. When they got to the hidden goal, the followers searched through the grass while Bandit simply stood and watched them. To some extent at least, chimpanzees can thus take into account each other's idiosyncrasies and fluctuations in motivation toward a given object.

To test the ability of two leaders to communicate the relative quantity of food in their respective hidden caches, we conducted a series of experiments in which we showed each a different pile of food (either large or small), returned them to the release cage, and turned them loose after a delay. In some control trials both animals were shown equivalent goals, so that the only way in which they could get to the large (preferred) pile first was to pool their information.

In one experiment in this series, Bandit and Belle were tested without any additional animals to serve as followers. Here, they almost invariably travelled together. On slightly more than 80% of the critical trials in which their food piles differed in size, they went to the larger goal first, regardless of who had seen it, and then went to the smaller goal, and shared both sets of rewards. Almost all 'errors' involved going to Belle's goal when it was small, a finding predicted from earlier tests on leadership hierarchies. In other words, leadership status and the preference value of food are additive variables determining group choice. On control trials where both animals were shown equivalent goals, they almost always went to Belle's goal first.

In other experiments, Belle and Bandit were tested in the same fashion, but were given six other animals as followers. Now they separated from each other on almost half of the trials, Belle going one way with one bunch of followers and Bandit going the other way or waiting at the release cage until he could recruit a following. The larger the piles of food, the more likely the leaders were to split. And, regardless of which leader had seen the larger of two piles of food, this goal attracted more animals than the small pile of food. The *group majority's* percentage of 'correct' responses here was almost 80%, and (as in the previous experiment) most of their 'errors' were due to their preference for Belle as leader.

f) Type of food. The same logic and the same procedures were used in examining whether the group knew which of two leaders had seen a preferred type of food (fruit rather than vegetables). Since the performance of the group was essentially the same there is no need to dwell on the details.

   g) Food vs. novel objects. Food and novel objects are both highly attractive incentives for chimpanzees, but they produce different approach reactions on the part of a group of chimpanzees. My chimpanzees rarely split into two separate parties going in opposite directions at the same time to get to two visible novel objects; but such splitting was commonplace for food. Also, most novel objects are approached more slowly than food, animals often decelerate instead of accelerating as they get to the object, and the first contacts are tentative if not cautious. Finally, food obviously disappears rapidly as soon as the animals get to it; whereas novel objects are not eaten and hence remain at least partially accessible to all for a longer period of time. With these facts in mind, and remembering the logic of our previous tests, we decided to test whether the group could discriminate these two classes of highly desirable objects, when the objects were invisible and one leader had seen one and the other leader had seen the other. STEWART HALPERIN, now at the Gombe Stream Research Center, played a major part in this experiment; he will report the full details in a separate publication.

   All possible six pairs of four leaders (Bandit, Belle, Bido, Gigi) were tested; and on a given trial each leader might be shown either fruit or a novel object. This yielded 24 different conditions, including the control conditions, trials on which both leaders saw identical goals.

   On the critical trials in which the goals differed, the group majority went to the predicted class of goal (food) first on 84% of the trials, regardless of the leaders involved. Of the few errors, most were produced by failure to follow Gigi, the least preferred leader. On trials where both goals were food, the group usually split. On trials where either one or both leaders had seen a novel object, the group split less than 40% of the time. Thus it was clear that the chimpanzees could convey information about these objects.

   h) A final note on 'what is communicated'. It should be clear that my data do not disclose exactly how many *independent* types of information about objects and events a leader is capable of conveying to his followers. It is conceivable that four factors – direction, relative degree of positive valence (incentive value), relative degree of negative valence, and the individual identity of the leader – are sufficient to account for all of the data. But this issue must be left open. The criterion I use for saying that the followers know the class or the quantity or the distance of an object is their differential response to situations varying in class of object alone, quantity of reward alone, etc.; it might be that the followers could not discriminate which type of variable was involved. Only further research on the precise properties of the effective (goal) stimulus can answer this question.

On the other hand, why would it be any more esoteric to say informally that the followers knew, for instance, the quantity of food at the goal from having interacted with the leader than to say that the leader himself knew the same thing from having seen the food directly? The operational criteria for assessing the content of a 'communication' are no different, and no more mysterious, than the operational criteria for assessing the content of any form of sign learning.

Assume, for the sake of argument, that the communication system of wild chimpanzees is so simple that the animals can without error convey to each other nothing more than the direction of a stimulus object to the nearest 10 degrees; the value of that object on a quantitative bipolar scale with five intervals (ranging from very undesirable, through neutral, to very desirable); and which of five individuals knows best the nature and direction of the object. This assumption is surely oversimplistic, and a minimal estimate. Assume, also, that there is approximately 20% loss of accuracy on complex messages involving all three types of information. (These numerical guesses are of course based on the data from my captive juveniles.) Even if these assumptions are granted, elementary calculations show that followers are capable of understanding about 720 alternative messages regarding goal objects (80% × 36 alternative directions × 5 object scale values × 5 potential leaders). And this total applies only to communication with respect to a single object on a single occasion; it does not take into account communication of the relative value of two or more goals, or moment-to-moment feedback and corrections of errors during the course of a longer sequence of actions. This figure compares not unfavorably with many human cases of nonverbal information processing with multidimensional stimuli [MILLER, 1956], not to mention the honeybee [HALDANE and SPURWAY, 1954].

Until such time as we know fully what chimpanzees are capable of doing with the communication systems they now possess without training on the part of human beings, it might well be premature to speculate as to how much more they might do if only they had somehow developed a few manual signals to indicate the distance, direction, and nature of distant goals [cf. HEWES, this symposium; LANCASTER, 1968]. Indeed, I wonder whether two chimpanzees who have been trained in humanoid sign language, like the GARDNERS' Washoe, but otherwise deprived of group-living experience with their own kind, could convey to *each other* any more information than group-living wild animals already do. The criterion of communication here must be not what their signalling responses look like to human observers, but the pragmatic consideration of what new inter-chimpanzee coordinations or what actions on the part of a follower can be accomplished by means of any signals. An 'untrained' control group is also essential; and the most informative comparison would be their respective joint pragmatic accomplishments, regardless of the form of their signals. I understand that several investigators are presently engaged in studying two-chimpanzee communication in language-trained animals; perhaps they will soon be able to answer these questions.

How Are the Animals Communicating?

To describe all of the types of signalling behavior that were observed in the course of this study would take a monograph by itself. Fortunately, other workers [e. g., VAN LAWICK-GOODALL, 1968; VAN HOOFF, 1967] have provided us with excellent descriptions of the chimpanzee's repertoire of signals and displays, which more than cover this problem. Therefore I shall concentrate on asking only what classes of signals seem to provide a sufficient basis for the observed communications about objects, as described in previous sections of this paper. I cannot pretend to know what role each class of signal in the repertoire of chimpanzees plays in the coordination of group life and in communication about objects. That would require at least a decade of further experimental research, and it is more than likely that my experiments have missed something.

Is it possible that the chimpanzees were communicating the direction and nature of an object while they were still in the release cage; e. g., by means of gestures? The behavior of these chimpanzees in the release cage, after we had informed the leader but before the group was released, was observed only occasionally. However, we saw nothing analogous to the 'dance language' of bees at the hive and nothing that could obviously account for the data. Most often, the leader wrestled with his companions or simply sat at the doorway until the group was released. In several experiments we placed the leaders and followers in separate release cages, with no effect on the group performance except that the leader whimpered and struggled to avoid being left alone. In one of these experiments, we first put the informed leader in the release cage with the rest of the group for several minutes (to permit exchange of information), then removed him, put him in a separate cage, and turned the followers loose to see if they could find the food. As we had feared, the followers did nothing, other than try to open the cage which held their whimpering 'leader'. When the leader was turned loose, however, all animals set off for the goal as usual.

These results do not, of course, disprove the possibility of communication within the release cage or prove that it was *only* what the leader did after release that was important for communication. They simply demonstrate once more the social dependency of the animals and the difficulty of disentangling the facts of object-communication from more general facts of social organization in primate groups. The experiments on communication of direction in which we served as 'surrogate leaders' were designed to overcome these difficulties.

In the course of our earliest experiments, we observed many instances in which the leaders tried to 'recruit' a following by means of glances, whimpering, arm signals, tapping a companion on the shoulder, presenting the back to solicit walking in tandem (with an arm around the waist), etc. The behaviors were both dramatic and convincing evidence to a new observer that the chimpanzees were indeed 'trying to communicate' and possessed a rich repertoire of potential ethological signals for doing so. The stark fact, however, is that two years later in the last months of research, by which time the chimpanzees were adolescent and showing their best performance in getting to the best available objects, we almost never saw these impressive displays. Indeed, even in the early tests it was Bandit, the most infantile and variable leader, who did the best and most frequent displays; and he showed them only on a fraction of his successful trials. In other words, the signals and displays that would enter into the typical observational account of communication were not *sufficient* to explain the experimental record of group performance. Moreover, the followers, to whom these displays were directed, seldom responded to them by getting up and following the leader. They waited until the leader oriented 'out there' and set off 'independently'. I would guess that young and inexperienced chimpanzees have a *richer* 'vocabulary' of humanoid-like signals than older and more experienced ones and that they are reinforced (perhaps by their companions' obtuseness) into abandoning them for something different. The success of the GARDNER'S experiment is probably due to the fact that only human beings value manual gestures and vocalizations for their own sake and can accordingly reinforce them 'appropriately'.

The qualitative summary notes that follow, written for the most part about two years before the chimpanzees were tested formally on communication, and four years before the termination of testing, suggest how this learning process might work. Van Lawick-Goodall [this symposium] has a surprisingly similar account of *mother-infant* 'training' in signalling; and Carpenter's [1964] work has parallel accounts on several other species.

July, 1967. Shadow, Bandit, Belle, and Louis have now been tested for 88 half-hour sessions in the field enclosure, and returned to their common home cage between tests. This note summarizes the most important changes in the ways in which a member of this close-knit group might initiate travel.

Especially in the early days when they were most cautious, they did a lot of ventro-ventral clinging and even tried at times to travel in this fashion. But more commonly they engaged in 'tandem walking', i.e., one animal walking behind the other, clinging to his waist. At first all four animals travelled in one long chain, but soon it was more common to see them in pairs. Also, instead of clinging tightly they grasped each other's hair only or simply draped an arm over a companion. (This tandem walking behavior is commonplace in wild-born infants and juveniles who are housed together, and it persists in some degree until the animals are adolescent. I believe it is a strict counterpart to a wild infant's riding on his mother's back.)

At first, whoever was most excited or whoever apparently wanted to initiate travel would grab onto the back of another; and if the initiator remained quadrupedal and pulled his companion into a standing position, his companion would start to walk. The result was often confusion, however. Sometimes the animal who was in front did not know where to go. Since in their home cage the animals seldom travelled more than a dozen meters in one direction, they also had not worked out the simple mechanics of walking well enough to keep from getting their legs tangled. And finally, if two or more animals (sometimes all four animals) were excited or upset, each would try to get in the rear position. The result was a lot of whimpering and shifting around and stumbling, until the initiator of travel took the lead position in the tandem chain.

Within a couple of weeks, a clear 'leadership' hierarchy started to appear. (This hierarchy was simply who walked in front; it bore no clear relationship to leadership in the ability to influence the direction of group travel, as measured in later tests.) Shadow went at the head of any chain and Louis was almost always at the rear. Belle and Bandit shared the middle 'ranks' (with Bandit slightly more apt to lead) and varied their position with respect to each other according to the circumstances. (Louis died not long after these observations; but the tandem-ranks of the other three remained much the same for four years, until tandem walking disappeared.) The chimps also eventually developed a well-synchronized march step so that they rarely got their legs tangled. The relationship between who initiated or steered the chain and who 'led' also disappeared. Any animal could 'steer' from any position. And, finally, the ways in which an animal initiated travel became increasingly more subtle. For example, Shadow came to simply hold Belle's back lightly and to raise her up into a quadrupedal position, and then take one step forward; at this she would take the rear position, and they would set off. There was no need to shift positions further.

Still later, after about a month, Shadow seldom did even this. Instead, he tapped Belle or Bandit on the shoulder, and he or she instantly stood up and got on his back. At that time this signal had never been observed to occur in the home cage; but once it had been developed in the large field enclosure, it transferred perfectly to the home cage.

By the 80th day of testing, all animals were using tapping as a signal to initiate tandem walking; and they also occasionally employed noncontact signals. Again, Shadow was the first to do so. He simply glances at a companion and then out into the field and 'presents' his back, and at this the others come to him. Also, on at least eight occasions he has extended his arm toward a companion, palm up, as if begging. Even if the others are several meters off, they come running. In several instances, however, Louis was climbing on the cage wire and did not respond to this begging gesture. Shadow reached up and touched his companion on the foot with a finger, then 'begged' again. Louis, who had been vigorously playing with a loose wire, instantly abandoned this activity, dropped down, and (as Shadow 'presented' his back) gave a little scream and got in rear tandem-position.

Vocalizations are of course a common way of eliciting contact from another. If any animal screams, others run to him, cling to him, and (a minute or so later) groom him. The louder he screams, the more companions come running. Whimpering is more apt to be followed by tandem walking or its nonlocomotor counterpart, ventro-dorsal clinging. However, the facial expressions preliminary to whimpering also suffice. For example, Bandit incessantly approaches Shadow with his lips puckered in a 'pout face', at which signal Shadow rises and 'presents' his back for tandem. If very upset, Bandit tries to climb on Shadow's back and ride dorsally like an infant; but Shadow collapses under the weight. It should be noted that the use of vocalizations and 'pouting' as a means of initiating tandem walking did not originate in the field enclosure tests; they had been used for at least a year in the home cage.

Probably the most impressive new form of coordinated travel and visual signalling, however, is the behavior we call 'pack running'. We have seen it only about a dozen times thus far. (It became more and more common in subsequent years, until it was at least a daily occurrence and the signals for it became so fleeting and subtle that all animals seemed to get the message 'independently' and simultaneously.) Most often it has occurred when the animals spot a novel object at a distance; but sometimes it occurs without any apparent goal other than just racing along together on an apparently pre-selected course. All animals stop and glance at each other and down the field in the same direction. Bandit might start moving for Shadow as if to get in tandem position, but Shadow turns so that he is *not* 'presenting' his back. Belle takes a few half-trotting steps out in front of the others, and at this Louis rolls ahead of her with a series of somersaults. All four chimps then set off almost shoulder-to-shoulder, but with no one in physical contact with another. First one animal and then another glances at his companions and steps up the pace just a little, until in a matter of 10 or 15 seconds the group has accelerated to top speed and is racing down the field. If a toy is the goal, they all dive for it. Otherwise they might reach the tree or the corner of the enclosure they had all apparently aimed for at the start, rest for a moment, and then start all over, glancing around for a new destination for the next run. If for any reason someone diverges from the apparently pre-selected route, or if there is some external distraction (e. g., when Louis spots a strange-looking rock just off the path), the 'pack running mood' seems to be broken immediately; and the more infantile tandem walking pattern is resumed.

These observations suggest that the smoothly-coordinated moves of free-living adults involve an enormous amount of learning not only of social signals, but also of the 'goal directed' quality of each other's behavior. The infant must in effect learn to extrapolate farther and farther ahead in space and time until eventually, as an adult, he need no longer worry about getting left behind, because he knows that when the group sets off they will continue in the same direction until they either hit a barrier or reach something worth stopping for. (The 'line of travel' need not be straight in the geometrical sense; it can be 'straight down that old pathway which twists all around the hill'.) The chief difference between my observations and those of field workers is that each of my chimpanzees seemed to play simultaneously the role of infant *and* the role of the mother or 'teacher'.

In my opinion, visual orientation and the locomotor postures and movements of the 'animal as a whole' contain sufficient information to account for the bulk of the observed communication about hidden objects. These 'signals' can easily provide a basis for directional communication. A quadrupedal chimpanzee orienting in a given direction is, if anything, a more accurate 'pointer', especially for an observer in a tree or on a laboratory tower, than a bipedal human being pointing in the same direction manually. One good reason that chimpanzees very seldom point manually is that *they do not have to;* rising to a quadrupedal position, glancing at a follower, and orienting 'out there' conveys all the directional information one could ask for [cf. SCHALLER, 1964]. HEWES [this symposium] argues, partly on the basis of the chimpanzee's ability to learn manual sign language, that human verbal language evolved from manual pointing gestures. But chimpanzees (if not nonprimate species such as dogs) can already communicate object-information through postural and locomotor pointing. Given the same cognitive ability, would it be *necessary* for the ancestors of man to pass through a manual stage also before switching to the vocal-auditory channel? I leave this question to the experts.

The presence of an object 'out there' could easily be conveyed by locomotor signals and glances, provided only that followers have learned through social experience that the attention of other living beings is usually elicited by and directed toward external events. The speed and, especially, sudden accelerations and decelerations of a leader's locomotion provides, further, a sufficient basis for judging the relative incentive value of the object.

This is not to say that more molecular signals such as piloerection, alarm calls, 'cautious hoots', and social tappings, tuggings, exchanges of visual glances, etc., do not provide additional information which supplements and reinforces that provided by gross signals. For example, it is doubtful that followers could discriminate whether a leader is running *toward* a highly desirable hidden object or *away from* a highly frightening one without some cues in addition to the leader's direction and speed of locomotion; and such a discrimination is probably elementary.

Nevertheless, from sample motion pictures of the animals as they emerged from the release cage we ourselves could usually predict within the first few seconds the direction of the best available object, which animals knew this direction, and whether or not some less preferred object was hidden in another direction. From a map of the simultaneous positions of all animals in the field at a given instant in time and the direction of their visual orientations, we could give essentially the same predictions, except that signs such as pilo-erection, etc., were useful additional information for determining whether the object was something desirable or something frightening.

More convincing from the standpoint of *chimpanzee* behavior were the tests on communication of direction in which we did not show the leader the hidden food, but only gave him a social signal and then removed ourselves from the situation. In one of these tests we gave the leader one of four cues on a given trial. As one experimenter held the chimpanzee just outside the release cage, another experimenter either: (a) walked a few paces in the direction of the hidden food, glanced back at the chimpanzee, and then leaned forward as if orienting toward something; (b) stood beside the chimpanzee and oriented and pointed manually in the direction of the hidden food; (c) walked a few paces as in (a) and then also pointed as in (b); or (d) turned slowly in a full circle while orienting all over, thus giving no reliable orientational information. After the cue had been given, the cue-giver returned to the front of the release cage, the chimpanzee was put back in the cage, and (several minutes later, with the experimenters gone) the group was turned loose.

Not surprisingly, the last, or control, condition led to either a 'random' search or no travel at all. In the other three conditions, the animal to whom we had given the cue struck off in the appropriate direction, usually with less than a 10-degree error; and his followers in turn got the cue from him. The crucial finding for our present purposes is, however, that there were essentially no differences in performance in these three conditions. Either oriented walking (which chimpanzees obviously do) or manual pointing (more strictly a human method of directional cuing) is a sufficient cue of direction, and the cue can be remembered even after the cue-giver has left his original position or disappeared from sight.

If anything, walking should be a better cue of location than any static pointing cue. The averaged angle of the path of travel contains less error (for an outside observer) than a more static cue; and also, if a leader has proceeded a given distance, the followers can be sure that the goal lies still further along this route. Thus, in other experiments we varied our cues to the leader by varying how far we walked in the direction of hidden food before returning to stand by the chimpanzee again. On release, the leader, followed by others, raced to the *end* of the path we had followed (sometimes taking a detour or shortcut to reach this point) and then walked slowly out in the appropriate direction (extrapolating ahead from our old path), visually searching for the food. This cannot be explained by simple trail-following, for on control trials in which we walked out to hide the food but gave no cue, performance was at a chance-level. In a few trials, we gave the chimpanzees a choice of two cues (and goals): a long walking cue in one direction, and a manual pointing cue in another direction. As predicted, the animals went in the former direction.

The experiment in which we varied the length of our cue-walk was designed on the basis of our observations of the chimpanzees themselves. If a leader travelled even two meters before he stopped and oriented back and waited for his followers, he seemed much more likely to obtain a following than if he stood at the release cage door and waited. Similarly, in the case of the two-leader tests, whichever animal started out the fastest and went the farthest before stopping to wait for followers (or returning to recruit them) stood the best chance of getting a majority following. If this should be the case, and if such variations are related to the nature of the goal object and the hierarchical status of the leader in the group, it is obvious that they provide a sufficient mechanism for communication. To gain control over the nature of the cues and to use them as independent variables in the experimental sense, we again used ourselves as 'surrogate leaders'. The chimpanzee leader was given a choice of two cues (and two goals); a long walk in a given direction signified a big pile of food, and a short walk signified a small pile of food. These two cues were given in succession, in varied orders. The four chimpanzees who were tested solved this 'problem', like nearly all of the others, on the first trial. They chose the longer walking-cue (and got the better goal) most of the time, regardless of the order in which the cue was given or the direction in which it was given.

Conclusions

The results of this study stand out in contrast to the conclusions which SCHALLER [1964] and KUMMER [1971] drew from their observational studies of leadership in gorillas and hamadryas baboons, respectively. SCHALLER says he saw no evidence that the leader himself had any definite goal or purpose in his travels, let alone that he communicated about these goals to his followers. KUMMER suggests that without symbolic language of a humanoid sort, primates cannot communicate to the followers about an object that is remote in space or time, except by taking them directly to it. The same opinions may be found in countless books and articles on human language and primate communication. Our chimpanzees in the experiments discussed here do not exactly fit this picture, and I do not believe that the species variable alone is operating. Two students at the State University of New York at Stony Brook, SAM GOLDBERGER and ROBERTA BLACK, recently tried some of the pointing experiments on human-animal communication with a golden retriever dog. The dog took several days to learn the task, but after this period the accuracy with which he went in the same direction as we had pointed out earlier matched the performance of my best chimpanzees. LORENZ's [1970] observations on jackdaws and MILLER and DOLLARD's [1941] laboratory studies of leadership in rats, and of course, many studies of foraging in ants and bees, suggest that similar or analogous forms of communication are operating at all phyletic levels. What is called for is clearly an experimental analysis of wild primate species, using methods comparable to those of other disciplines but suited to the social organization and foraging patterns of the species at hand. And, once the fact of primate 'communication about the environment' is accepted as ubiquitous, we can get on to more interesting problems, such as how it develops and functions in a given society and how the mechanisms of non-human primate communication about the environment differ from those of ants or men.

Although I have laid great stress on the role of rather gross locomotor and postural cues in chimpanzee communication, the data do not, of course, disprove the value of more molecular studies of signalling behavior (gestures, facial expressions, vocalizations, and so on). Rather, they suggest that to understand the communicatory significance of ethological displays, one must place them into their larger context. Specifically, a 'complete' analysis of visual communication patterns, in terms of overtly detectable motion, should take into account all three independent mechanically-defined 'levels' of motion: (a) the movement of the animal's hands, eyes, and other body parts with respect to his center of gravity; (b) the movement of this center of gravity point ('the animal as a whole') with respect to the spatial centroid of his social group; and (c) the movement of this social centroid with respect to some external reference point. The first level (at which many ethologists and behaviorists claim to be working) would, if taken *alone*, explain only a small fraction of social communication. But, of course, neither animals nor men attend only to each other's limb movements. Ordinarily, to get to food and escape danger, all a group-living primate has to do is follow the horde (level c). In cases of uncertainty or conflict, however, the movement of a preferred companion or a high ranking individual with respect to the group's general line of travel (level b) provides additional information. And if this still does not suffice, the more molecular signals of this individual (level a) decrease the total uncertainty of the situation still further.

There is no good reason that the above sequence cannot be reversed, as when a single alarm call galvanizes a whole group into flight, and so perhaps it is not so much that animal displays lack all 'propositional' quality as that animals are more dependent than man on 'composite signals' and contextual factors to convey a message [cf. MARLER, 1965; COUNT, this symposium]. For chimpanzees, if not for many 'lower' species, locomotion conveys at least some of the object-information which in our more highly evolved and differentiated species is customarily conveyed by gesture and speech. LOGAN [1972] in fact presents considerable evidence that variations in the speed of locomotion toward different incentives are learned according to the outcomes they produce, rather than being innately determined 'strengths' of a single response. On such a theory, group-living animals could learn to use each given speed or type of locomotion as a social code for controlling the group in a particular fashion, in exactly the same fashion that discrete gestures or vocalizations might be used for this purpose. Quite possibly, stylized displays evolved from more primitive locomotor 'intention movements' [LORENZ, 1970], but thereafter became more inflexible and less subject to operant learning.

Finally, these data would seem to support the GARDNER and PREMACK studies, which show that chimpanzee as a species is capable of learning something which possesses most of the 'design features' of language. Why would such a capacity have evolved if it served no function for the feral ancestors of laboratory chimpanzees? And is it not likely that wild chimpanzees are exploiting some aspects of that capacity even today? In my opinion, animals and man share a basic communication system which goes much deeper than gesture or other superficial similarities in the form of their ethological displays. To call it 'language' does not necessarily add to our understanding. However, it does serve the same biological function as language – namely, to coordinate and regulate social life, and to achieve common, species-specific environmental goals. The proper place to look for it is not in the display systems *per se*, 'but rather in systems for organization of perception or something of that type' [CHOMSKY, 1967; see also COUNT and EISENBERG, this symposium].

References

ALTMANN, S.: A field study of the sociobiology of rhesus monkeys, *Macaca mulatta*. Ann. N.Y. Acad. Sci. *102:* 338–435 (1962).

ALTMANN, S.: The structure of primate social communication; in ALTMANN Social communication among primates, pp. 325–362 (University of Chicago Press, Chicago 1967).

ATTNEAVE, F.: Applications of information theory to psychology (Holt, Rinehart & Winston, New York 1959).

CARPENTER, C. R.: Naturalistic behavior of nonhuman primates (Pennsylvania State University Press, University Park 1964).

CASSIRER, E.: An essay on man (Yale University Press, New Haven 1944).

CHOMSKY, N.: The general properties of language; in DARLEY Brain mechanisms underlying speech and language, pp. 73–88 (Grune & Stratton, New York 1967).

CRAWFORD, M. P.: Further study of cooperative behavior in chimpanzees. Psychol. Bull. *33:* 809 (1936).

CROOK, J. H.: The socio-ecology of primates; in CROOK Social behaviour of birds and mammals, pp. 103–166 (Academic Press, New York 1970).

DAVENPORT, R. K.; ROGERS, C. M., and MENZEL, E. W.: Intellectual performance of differentially reared chimpanzees. III. Discrimination learning set. Amer. J. ment. Defic. *73:* 963–969 (1969).

FRISCH, K. VON: The dancing bees (transl. ILSE) (Harcourt, Brace, New York 1955).

GARDNER, B. T. and GARDNER R. A.: Two-way communication with an infant chimpanzee; in SCHRIER and STOLLNITZ Behavior of nonhuman primates, vol. 4, pp. 117–184 (Academic Press, New York 1971).

GIBBS, C. A.: Leadership; in LINDZEY and ARONSON The handbook of social psychology; 2nd ed., vol. 4, pp. 205–282 (Addison-Wesley, Reading 1969).

GOODALL, J.: Chimpanzees of the Gombe Stream Reserve; in DEVORE Primate behavior, pp. 425–473 (Holt, Rinehart & Winston, New York 1965).

HALDANE, J. B. S. and SPURWAY, H.: A statistical analysis of communication in *Apis mellifera* and a comparison with communication in other animals. Insectes soc. *1:* 247–283 (1954).

HALL, E. T.: The silent language (Doubleday, Garden City 1959).

HAYES, K. J. and HAYES, C.: The cultural capacity of chimpanzee. Human Biol. *26:* 288–303 (1954).

HEBB, D. O. and THOMPSON, W. R.: The social significance of animal studies; in LINDZEY Handbook of social psychology, vol. 1, pp. 532–561 (Addison-Wesley, Cambridge 1954).

HOCKETT, C. F.: Logical considerations in the study of animal communication; in LANYON and TAVOLGA Animal sounds and communication, pp. 392–430 (American Institute of Biological Sciences, Washington 1960).

HOOFF, J. A. R. A. M. VAN: The facial displays of the catarrhine monkeys and apes; in MORRIS Primate ethology, pp. 7–68 (Weidenfeld & Nicholson, London 1967).

ITANI, J.: Vocal communication of the wild Japanese monkey. Primates *4:* 11–66 (1963).

ITANI, J. and SUZUKI, A.: The social unit of chimpanzees. Primates *8:* 355–381 (1967).

KÖHLER, W.: The mentality of apes (transl. WINTER) (Harcourt, Brace & World, New York 1925).

KUMMER, H.: Social organization of hamadryas baboons (Karger, Basel 1968).

KUMMER, H.: Primate societies (Aldine-Atherton, Chicago 1971).

LANCASTER, J.B.: Primate communication systems and the emergence of human language; in JAY Primates: studies in adaptation and variability, pp.439–457 (Holt, Rinehart & Winston, New York 1968).

LAWICK-GOODALL, J. VAN: The behaviour of free living chimpanzees in the Gombe Stream Reserve. Anim. Behav. Monogr., vol. 1/3 (Ballière, Tindall & Cassell, London 1968).

LAWICK-GOODALL, J. VAN: In the shadow of man (Houghton Mifflin, Boston 1971).

LOGAN, F.: Experimental psychology of animal learning and now. Amer.Psychol. *27:* 1055–1062 (1972).

LORENZ, K.: Studies in animal and human behaviour, vol. 1 (Harvard University Press, Cambridge 1970).

LYONS, J.: Human language; in HINDE Non-verbal communication, pp.49–85 (Cambridge University Press, Cambridge 1972).

MACKAY, D.M.: Formal analysis of communicative processes; in HINDE Non-verbal communication, pp. 3–26 (Cambridge University Press, Cambridge 1972).

MARLER, P.: Communication in monkeys and apes; in DEVORE Primate behavior, pp. 544–584 (Holt, Rinehart & Winston, New York 1965).

MARLER, P.: Aggregation and dispersal: Two functions in primate communication; in JAY Primates: studies in adaptation and variability, pp.420–438 (Holt, Rinehart & Winston 1968).

MENZEL, E.W.: Responsiveness to objects in free-ranging Japanese monkeys. Behaviour *26:* 130–150 (1966).

MENZEL, E.W.: Responsiveness to food and signs of food in chimpanzees' discrimination learning. J. comp. physiol. Psychol. *68:* 484–489 (1969).

MENZEL, E.W.: Group behavior in young chimpanzees: responsiveness to cumulative novel changes in a large outdoor enclosure. J. comp. physiol. Psychol. *74:* 46–51 (1971a).

MENZEL, E.W.: Communication about the environment in a group of young chimpanzees. Folia primat. *15:* 220–232 (1971b).

MENZEL, E.W.: A group of young chimpanzees; in SCHRIER and STOLLNITZ Behavior of nonhuman primates, vol. 5 (Academic Press, New York, in press).

MENZEL, E.W. and DAVENPORT, R.K.: The effects of stimulus presentation variables upon chimpanzees' selection of food by size. J. comp. physiol. Psychol. *55:* 235–239 (1962).

MILLER, G.A.: The magical number seven plus or minus two: some limits in our capacity for processing information. Psychol. Rev. *63:* 81–97 (1956).

MILLER, N.E. and DOLLARD, J.: Social learning and imitation (Yale University Press, New Haven 1941).

MORRIS, C.: Foundations of the theory of signs; in NEURATH, CARNAP and MORRIS International encyclopedia of unified science, pp. 77–137 (Chicago University Press, Chicago 1955).

MOWRER, O.H.: Learning theory and the symbolic processes (Wiley, New York 1960).

PAVLOV, I.P.: Conditioned reflexes (Oxford University Press, Oxford 1927).

PREMACK, D.: On the assessment of language competence in the chimpanzee; in SCHRIER and STOLLNITZ Behavior of nonhuman primates, vol.4, pp.185–228 (Academic Press, New York 1971).

PREMACK, D.: Concordant preferences as a precondition for affective but not symbolic communication (or How to do experimental anthropology). Presentation, Conf. Behavioral Basis of Mental Health, Galway 1972.

REYNOLDS, V. and REYNOLDS, F.: Chimpanzees of the Budongo Forest; in DEVORE Primate behavior: field studies of monkeys and apes, pp. 368–424 (Holt, Rinehart & Winston, New York 1965).

SCHALLER, G. B.: The year of the gorilla (University of Chicago Press, Chicago 1964).

THOMPSON, R.: Approach versus avoidance in an ambiguous-cue discrimination learning problem. J. comp. physiol. Psychol. *47:* 133–135 (1954).

TINKLEPAUGH, O. L.: An experimental study of representative factors in monkeys. J. comp. Psychol. *8:* 197–236 (1928).

TOLMAN, E. C.: The determiners of behavior at a choice point. Psychol. Rev. *45:* 1–41 (1938).

WIENER, N.: Cybernetics (MIT Press, Cambridge 1961).

Author's address: Dr. E. W. MENZEL, jr., Department of Psychology, State University of New York at Stony Brook, *Stony Brook, NY 11790* (USA)

Symp. IVth Int. Congr. Primat., vol. 1: Precultural Primate Behavior,
pp. 226–231 (Karger, Basel 1973)

Dominance versus Possession

An Experiment on Hamadryas Baboons

H. KUMMER

University of Zürich

A living system requires co-ordinating mechanisms which insure that no particular activity is carried out by too many or too few of its parts. One solution, realized among tissues of a metazoan organism, is to differentiate firmly and permanently the response patterns of the parts at an early stage of the system's ontogeny. Vertebrate and most invertebrate societies are not organized on this principle. All members of a species develop remarkably similar response patterns and maintain them throughout their lives. They are not programmed to perform only a few types of activities; rather, every adult member is basically prepared or is even actively seeking to perform all activities occurring in the society. All adult primates have a tendency to mother infants, to scan the surrounding landscape, and so on. What varies with age, sex, and individuality is not the presence or absence of a given behavior, but the frequency of the behavior and the circumstances under which it occurs.

The flexibility of a society composed of pluri-potential members has obvious advantages in semi-permanent or small groups, which must frequently reorganize, e.g. after a member has died. On the other hand, such a society must be capable of *ad hoc* differentiation; it needs mechanisms that inhibit some members and disinhibit others in situations that require differential courses of action whenever individual difference is an insufficient differentiatior. In a group of hunting dogs, for example, one adult remains with the pups while the others go hunting; and this differentiation takes place anew before every hunt [KÜHME, 1965]. In practice, most societies do not differentiate afresh in every emerging situation. Instead, they show various types of semi-permanent, reversible differentiations that are learned by performers and non-performers alike. Roles emerge in which certain individuals habitually perform particular activities in most recurrent situations of a particular kind.

The most conspicuous principle of semi-permanent differentiation is dominance. In the light of the above deductions it is striking that in its classical definition, a high dominance rank is not so much a role that benefits the society but one that, in its immediate effect, serves no one but the bearer. It has been reported that in certain species the dominant primate also signals the location of his group [GAUTIER, 1969], leads the way [SCHALLER, 1963] or defends the group [DEVORE and HALL, 1965], but it seems that in several cases these altruistic activities are attributes of the adult male rather than the dominant animal. Where there is more than one male, the tasks of exploration, leadership, or defense are quite often carried by subordinates. It is also probable that priority rights within the group may free the dominant animals for such tasks as gathering and processing information on spatially remote events. In wolves, for example, the dominant animals frequently orient at distant stimuli; whereas subordinates rarely do so [ZIMEN, 1971]. These considerations do not answer the questions: How do the subordinates survive? How do they secure sufficient food or safe resting places when and where such things are scarce? The answer may be that they are in fact disadvantaged under such conditions and that selection takes its course, checked to some degree by cooperative and protective behavior from dominant animals.

Qualitative field evidence strongly suggests that this explanation is not the complete answer. Even where two individuals seem to compete for a highly desired, rare, and spatially concentrated incentive, the outcome does not always fit the dominance model. A large female can be dominant over a smaller female with respect to food, but she is most probably subordinate when attempting to approach the latter's infant. Such inconsistencies of a dominance order could still be explained by demonstrating that one female is more strongly motivated to take food, whereas the other is a more eager infant-handler. But inconsistencies appear even where they are clearly not caused by different motivations. When a piece of food is thrown midway between two adult male baboons, the more dominant will usually take it. If, however, the food falls one meter from the subordinate and ten meters from the dominant, there is a good chance that the former will take it without reprisals. When a subordinate chimpanzee is eating a large piece of meat, a dominant chimp either 'ignores' it; or, if he approaches, he may actually have to beg for a share [VAN LAWICK-GOODALL, 1968]. According to MENZEL [personal communication], dominant chimpanzees at the Delta Regional Primate Research Center usually proceeded with wile rather than force when they attempted to steal a toy from a subordinate; they distracted him by playing, then snatched the toy. Often not only the deprived chimp but others as well pursued and threatened the 'thief'. An object seemed to 'belong' to an individual at least until he put it down completely, although the more dominant the 'owner', the greater distance he could move away before others went for it.

Similar results were obtained in a study on hamadryas baboons [KUMMER *et al.*, in press]. In this case, the competing animals were adult males and the 'object' of competition was an adult female of the same species. Hamadryas males are known to lead and herd exclusive harems of females and to fight violently every rival male that attempts to interact with them. When the occasion arises, many males eagerly enlarge their harems [KUMMER, 1968]. Under these conditions, one would expect either a pronounced dominance order or a high rate of fighting among the males of a troop. In reality, fighting is rare; and an analysis of one band showed a low correlation between a male's ability to displace another at a baiting place and the number of his females. Several lines of evidence presented in the original 1968 publication led to the impression that dominance is not the only principle of female distribution. It may decide which male will secure an alien female that is artificially released into a troop by an experimenter; but once a male is associated with her, he will usually be able to keep her even against a dominant rival. We hypothesized that a male's tendency to acquire an additional female is inhibited by seeing her interact with another troop male, regardless of their dominance relationship.

The experiments reported in this paper were carried out in a wire net enclosure on our camp grounds in order to prevent any one-sided advantage that a male might gain from being on familiar ground or among familiar troop members. Two adult males from each of four troops and 15 adult females from various troops were trapped and habituated to the enclosure. In every trial, two of these males and one of these females (that neither male had seen before) were released into the enclosure. The sequence of admission was systematically varied in a total of 39 trials. The series most relevant to the present context consisted of 16 trials in which a male, designated as 'the owner', was admitted to the enclosure and given a female; and the other male from his troop, called 'the rival', was then allowed to watch the pair for a 15-minute exposure period from a large wire cage outside the enclosure, whereupon the three baboons were united in the enclosure. (Possible cage effects were balanced by placing the pair in the cage and the rival in the enclosure in half the trials.) Each pair of males from the same troop went through four trials together, every time with a new female. Each male served alternately as the owner and as the rival. If dominance alone had determined the outcome, the female would have fallen to the same (i.e. the dominant) male in every trial, whether he started out as owner or rival. In reality, in this series of trials the female always remained with the owner; and no rival attempted to take her by force. The dominance hypothesis was rejected at the 0.01 level of significance, even though we assumed that uninhibited dominance would have been effective in transferring the female in only one-half of all trials. During the trials, the rivals often stayed rigid, scratched with an abnormally high frequency, groomed themselves, and inspected 'irrelevant' parts of the environment. The ethological analysis showed that both their aggressive and associative behaviors were inhibited. We concluded that hamadryas pair bonds are protected by an inhibition in the rivals. It is not the more dominant male who keeps the female, but the male who had her first.

From the examples given, it seems that the subordinate primate survives not just because dominance is inefficient in a haphazard way. Rather, dominance is moderated and restricted by a principle that consistently favors the early bird. How does this principle work?

First, the 'early bird principle' essentially involves the inhibition of a competitor by a perceived relationship between a subject and an object. In the baboon example, it is known that the rival would freely interact with the owner or with the female if he encountered either one alone. What inhibits his interactions with the two together is a pair gestalt, which provides stimulus effects that differ from the summed effects of the two members. The effective gestalt can be roughly inferred from our experiments. The baboon females usually stayed less than 0.5 meters from the owners. This proximity and one or two casual interactions between owner and female were sufficient to inhibit the rivals. In the toy-handling and meat-eating chimps, it is again the spatial proximity and the tactile interaction between subject and object that inhibit the competitor. The example of food thrown close to a subordinate individual suggests that proximity alone may produce the inhibiting pair gestalt. The term 'personal space' seems to describe essentially the same phenomenon by focusing on the area within which an object appears to be related to the individual in the center.

Secondly, the triadic situation of a subject, an object, and a competitor may not only inhibit the latter. It may also disinhibit the subject, as was demonstrated by our baboons. When the pair could see the rival, the owner clasped the female significantly more often and the pair groomed more than when the rival was hidden behind a one-way screen. Furthermore, an inhibiting subject-object relationship can be established to the benefit of a subordinate subject only if he is initially favored by unequal conditions. One would expect that subordinates have techniques to create such conditions, such as the 'initial group' technique of hamadryas baboons in which young subordinate males acquire juvenile females before the latter are sexually attractive. The early-bird principle protects these pair bonds even when the female becomes an incentive for the dominant troop males. Finally, first establishing relationships with an object moderates the effects of dominance; but it does not always entirely prevent them. A different series of trials in the hamadryas experiments showed that very dominant rivals sometimes attack very subordinate owners and take over the females.

In human language, the early bird phenomenon is adequately described by the term 'possession', the meaning of which was originally to sit in power and, more specifically, is 'the recognized power of immediate interaction with an object'. It is worth noting that, according to the preliminary evidence, possession of a piece of dead matter and of a social partner seem to be recognized on the basis of the same kind of relationship (proximity and interaction). Given the similarity, it is appropriate to extend future investigations to all types of objects and to gain the advantage of comparisons. One difference is already obvious. At least in baboons, two group members can 'own' one another, each defending the relationship against competitors of his own sex-age class.

Whether possession and its complement, the competitor's inhibition, are precultural in the sense of this Symposium is an open question. In my present view, preculture consists of behavioral modifications (in the biological definition of the term) induced by the same behavioral modification of conspecifics. One more criterion seems appropriate: the biological function of modifications in general, which is to permit adaptive change within one generation, is only possible if a gene pool is capable of at least two alternative modifications (e.g. the presence or absence of a trait), each of which is adaptive within a certain range of socio-ecological conditions. If a gene pool were capable of only one viable modification, precultures and cultures could not undergo the process of historic change that is in fact observed. Therefore, if possession as such is in fact a necessity in every viable, free-ranging primate society, whatever its living conditions, I would not consider it as an element of preculture. Precultural variation probably exists, however, in that some groups are more possessive of infants; others are more possessive of female consorts or of a kind of food. Such variants could conceivably be transmitted by social modification; seeing adults interact intensely with a class of objects could orient the interest and possessive motivation toward such objects in the onlooking juveniles; and the competitors' inhibition could be learned by punishment.

Precultural forms of possession must of course be distinguished from property, which is a truly cultural phenomenon. Property relationships are not defined exclusively in terms of proximity and interaction; one may own an object as property but not possess it, and vice versa.

Summary

The available evidence suggests that dominance is not a complete or self-sufficient principle governing the order of access to objects. It determines priority with respect to entire classes of objects, if conditions are similar for both competitors. Possession, on the other hand, determines priority with respect to a single object. It can be established by a subordinate only when the initial conditions are sufficiently unequal. If these statements are correct, possession increases the chances of access for the subordinate. It also benefits a dominant animal when its vigor is temporarily reduced, e.g., by disease. When the object is a social partner, possession will protect pair bonds and thus increase the stability of a social structure.

References

DeVore, I. and Hall, K. R. L.: Baboon ecology; in DeVore Primate behavior, pp. 20–52 (Holt, Rinehart & Winston, New York 1965).
Gautier, J.-P.: Emissions sonores d'espacement et de ralliement par deux cercopithèques arboricoles. Biol. gabon. *5:* 117–145 (1969).
Kühme, W.: Freilandstudien zur Soziologie des Hyänenhundes (*Lycaon pictus lupinus* Thomas 1902). Z. Tierpsychol. *22:* 495–541 (1965).
Kummer, H.: Social organization of hamadryas baboons (Karger, Basel 1968).
Kummer, H.; Götz, W., and Angst, W.: Triadic differentiation: an inhibitory process protecting pair bonds in baboons (in press).
Lawick-Goodall, J. van: The behaviour of free-living chimpanzees in the Gombe Stream Reserve. Anim. Behav. Monogr., vol. 1/3 (Ballière, Tindall & Cassell, London 1968).
Schaller, G. B.: The mountain gorilla (University of Chicago Press, Chicago 1963).
Zimen, E.: Wölfe und Königspudel. Vergleichende Verhaltensbeobachtungen, p. 109 (Piper, München 1971).

Author's address: Dr. Hans Kummer, Institute of Zoology, University of Zürich, Aussenstation, Birchstrasse 95, *CH-8050 Zürich* (Switzerland)

Symp. IVth Int. Congr. Primat., vol. 1: Precultural Primate Behavior,
pp. 232–249 (Karger, Basel 1973)

Mammalian Social Systems:
Are Primate Social Systems Unique?

J. F. EISENBERG

National Zoological Park, Smithsonian Institution, Washington, D.C.

Introduction

One of the purposes for writing this brief review was to respond to a
suggestion from the Symposium Chairman. He hoped that I could place the
burgeoning studies of primate social systems into some perspective by relating primate societies and their attributes to the social systems of other mammals. I will attempt to do this, but ask you to bear in mind that the class
Mammalia, although maintaining some prominence in the terrestrial ecosystem, is in reality only a very small segment in the diverse spectrum of animal
forms. If the studies of primate social systems are to be considered a narrow
specialization, surely the study of mammalian social systems is only a trifle
less narrow. To place primate social systems in a biological perspective would
require a symposium in itself and we would have to be introduced to another
phylum, the Arthropoda. The members of this phylum have attained degrees
of social organization without parallel in the phylum Chordata, which, of
course, includes the class *Mammalia*. One could well contemplate the phenomena of caste differentiation, cohesion, and the whole concept of altruistic
behavior by an appreciative look at the termites, wasps, ants and bees (orders
Isoptera and Hymenoptera), as so elegantly portrayed in E. O. Wilson's recent
volume [WILSON, 1971]. With this brief homage to other fields of endeavor in
zoological research, let me return to the confines of the class Mammalia and
consider a few salient points.

Since one of the questions that this Symposium has considered turns
upon a definition of culture, let me indicate where I stand with respect to the
concept of culture. The term 'culture' became refined within the discipline of
anthropology; thus, I accept the notion that: 'culture includes those objects

or tools, attitudes and forms of behavior whose use under given conditions is sanctioned by the members of a particular society ... It is axiomatic that cultural values cannot be transmitted through biological inheritance so that each child must start to learn the culture of its group after it has been born' [TITIEV, 1955]; or, as an expansion of the term, (a) 'the act of developing the intellectual and moral faculties, especially by education'; or (b) 'the characteristic features of a particular stage of advancement in civilization' (Webster's Seventh New Collegiate Dictionary, definitions 2 and 5b).

As I understand it then, culture is an anthropological term most properly applied to the activities of mankind. The human brain apparently endows him with the capacity to assign meanings to objects or utterances which in and of themselves do not immediately denote or connote anything specific[1]. This capacity to assign meaning to objects which have no apparent meaning in themselves is generally referred to as the capacity for symbol formation, and as such is an aspect of culture. The ability to manipulate symbols and create an abstract communication system based on symbolic combinations certainly sets man's communication patterns apart from those of other animals. It is probably true, however, that man has a finite set of combinatorials or modes of combining symbolic expression [COUNT, this volume]. Indeed it has been suggested that man is not only preprogrammed for symbolic thinking within certain formal constraints but that these constraints can be described as the 'deep syntactical structures' in a human cognitive communicatory network [CHOMSKY, 1965]. Man's construction of artifacts as well as his ability to form a complex of symbolic communication based on learning and passed on from one generation to the next is unique, but the roots of such a system can certainly be traced and discerned in other mammalian species. Sophisticated analyses of human behavior and their roots in infrahuman mammalian forms have been pioneered by LORENZ [1971].

The establishment of continuities between *Homo sapiens* and the extant Pongidae is an over-riding preoccupation in sociobiology [JAYNES and BRESLER, 1971; DEVORE, 1971]. If we grant that symbolic communication of a very sophisticated order is the framework upon which culture hangs [R. FOX, 1971], then one task is to seek evidence of symbolic communication in *Pan, Gorilla, Pongo*, etc. Evidence concerning the ability of *Pan* to communicate information about objects and events in the environment [MENZEL, 1971, 1972] is one step in this direction. The ability of *Pan* to learn a symbolic mode of communication (based on gestures) and further to learn the rudi-

1   Of course the roots of some if not all aspects of symbol formation may lie before the development of speech as we know it today. [PAGET, 1930]

ments of syntactical rules is a step toward finding the limits of one aspect of primate learning [GARDNER and GARDNER, 1969; HEWES, this volume]. Such attempts at establishing continuities must, however, be based on a firm understanding of the communication system that functions in a free-living troop of chimpanzees. To this end pioneer efforts in the analyses of gestural and auditory communication in a free-ranging group must be acknowledged and noted [MARLER, in press].

Continuities may also be built in reverse by inspecting the stimulus preferences and exploratory behavior in a variety of vertebrates and then relating the stimulus pattern preferences to the human experience. In other words to explore the origin of the esthetic sense [RENSCH, 1958; MORRIS, 1962]. Tool use in mammals and birds may be profitably explored in the same manner and has been reviewed by VAN LAWICK-GOODALL [1970]. From this research we know that many mammals possess the capacity to manipulate objects. Indeed, objects in their environment can be modified; and, in the case of JANE GOODALL's chimpanzees, they can be modified apparently with a purpose of future use in mind. The example of the chimpanzee termite-extracting technique conforms to the rudiments of tool-making. Although many species of mammals employ sticks, bones, branches, and other artifacts in assisting them to manipulate their environment, only in the case of the chimpanzee, and perhaps the elephant *(Elephas maximus)* do we have reasonable documentation that an object is actually modified in such a way as to indicate that the maker (elephant or chimpanzee) has in mind a form to which the tool should conform as construction proceeds [McKAY, 1973; VAN LAWICK-GOODALL, 1970].

One aspect of all this research is the acknowledgement of the difficulty if not impossibility of drawing a sharp boundary between the behavior of *Homo* and other mammals. RENSCH [this volume] finds continuities between man, primates, and lower vertebrates; VAN LAWICK-GOODALL [1970] finds tool utilization in birds and mammals; MENZEL [1971] suggests communication about the environment is not the prerogative of higher primates alone and cautions against the use of poorly defined or all too readily accepted assumptions in the field of animal communication.

Leaving aside artifact construction and inferred symbolic processes then, is there anything that sets higher primate societies apart from the social organizations formed by other mammalian species? Here, I think, what appears to set primate societies apart from species of other mammals are sets of attributes characteristic of rather long-lived, big-brained mammals that have the capacity for the storage and retrieval of a great deal of individually-acquired information about the environment. I would like to consider four examples of erroneous preconception concerning the behavior of primates which are current in primatological research. These are the claims (1) that the behavioral repertoire of higher primates is more complex than that shown by more 'primitive' mammals; (2) that communication systems of higher primates are more complex than those of nonprimates; (3) that higher primates exhibit learned traditions and this ability sets them apart from the behavior patterns of other mammalian species; and (4) that higher primate social organizations are infinitely more complex than those of 'lower' mammals.

The Complexity of Repertoires

An ethogram or behavioral inventory for a mammalian species generally includes some 70 to 100 postures, movement sequences, vocalizations, and interaction patterns. Ultimately the length of the list depends on the degree of subdivision the author elects to make when analyzing a behavioral sequence into its components. Often the ethologist uses a functional classification in formulating a list with two major subdivisions: (a) the behavior of the solitary animal and (b) the behavior of interacting individuals. Because of this division, there is often about a 20-percent overlap between the elements of the two lists based on the two somewhat arbitrary (social versus solitary) conditions. Thus, absolute counts of behavior patterns from an ethogram tend to run slightly higher than the actual motor repertoire of the animal itself. For example, a list of some 119 terms, exclusive of vocalizations, will serve to describe the behavior patterns for most muroid rodents; 45 terms from the list would include the behavioral units employed in sexual behavior, maternal care, contact promoting, and agonistic behavior. If one includes some eight basic sound patterns, then 53 terms are generally adequate to cover the social behavior of a wide variety of rodent species [EIBL-EIBESFELDT, 1958; EISENBERG, 1967, 1968]. ALTMANN [1962] employed a list of 56 patterns to describe the social behavior of *Macaca mulatta;* this list included vocalizations.

WEMMER [1972], utilizing a very objective set of criteria, described the behavior patterns for the semi-solitary viverrid carnivore *Genetta tigrina.* Some 38 patterns served to categorize the behavioral units shown by a solitary animal; however, many of the patterns are shown in distinctive combinations thus increasing the actual list. Interaction patterns of associated animals may be described by utilizing 40 motor units and vocalizations. This list includes scent emission and scent deposition in the total repertoire. Once again, it should be noted that the basic list of movements, vocalizations, and scent emission patterns show several defined combinations; thus, if the combinatorials are added into the list as elements in their own right, the behavioral inventory for the genet may reach some 76 items. ALTMANN [1962] encountered a similar problem with *Macaca mulatta*, noting that 27 common combinations were formed out of his basic list of 56 units, thus expanding his total listing to 83 elements and combinatorials.

The point of this relatively brief review is not only to point out the difficulties involved in a comparison of mammalian repertoires but also to underscore the fact that most repertoires are at the same levels of apparent complexity whether or not the species lives in a social group (e.g., *Genetta* versus *Macaca*). Either this situation reflects an inherent limitation on the part of the human investigator to categorize behavior, or primate repertoires are not necessarily any more complicated than the behavioral repertoire of a mouse or a small carnivore.

Perhaps a way of illustrating the foregoing conclusion is to discuss one aspect of social analysis, the two-animal encounter. This is an artificially staged encounter where two animals (and for simplification, we will consider only male-female combinations here) are simultaneously introduced into a living space in which they have not established 'territorial ownership' [for details, see EISENBERG, 1969]. Let us take two extreme examples, an insectivorous mammal, the wandering shrew *(Sorex vagrans)*, and the New World primate, *Ateles geoffroyi*, both of which have been studied in some detail [EISENBERG, 1964; EISENBERG and KUEHN, 1966].

The shrew shows little reliance on vision, but its reliance on audition during an encounter is marked [GOULD, 1969]. In fact, shrews are characterized by an almost continuous emission of vocal signals that change in frequency and pitch depending on proximity to a conspecific. One of the attributes of their encounter behavior is to avoid one another without a direct approach; hence, if one tabulates *moving away* as a behavior pattern, it will show a preponderance not corresponding to a category *moving toward*, since the animals can avoid one another without a directed approach on the part of a partner. Chemical communication is very important in shrew encounters, and glandular secretions from specialized areas on the body are employed in marking and in emitting scent into the air.

In an encounter one generally finds that the shrews move about avoiding contact with one another initially; and when they have become slightly familiar with the environment, one shrew will attempt to approach the other. This can lead to touching noses and vocalizing; or if one animal starts to move away slowly, the second may follow and initiate a nose to genital or anal con-

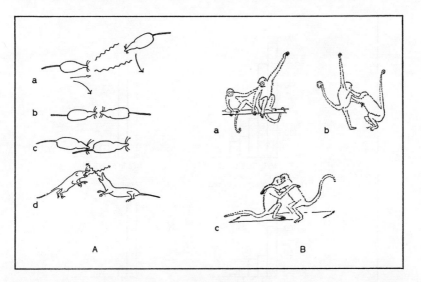

Fig.1. A Typical interaction sequences in an encounter between a male and female *Sorex vagrans.* (a) Approaching animal emitting ultrasonic pulses induces moving away in the partner. (b) Approach with nose contact. (c) First approaches second and sniffs the anogenital area. This may be followed by moving away on the part of the receiving animal or flight. (d) Mutual upright vocalizing intensely. This is a form of threat behavior. *B* Some interaction patterns for *Ateles geoffroyi.* (a) Touching. (b) Grappling. (c) Embracing.

figuration. Agonistic behavior rarely leads to a fight, but frequently involves rearing on the hind legs, mutual uprights, and vocalization (fig. 1). If a female shrew is receptive to a male, the subsequent reaction grades into driving where the male persistently follows the female and eventually she will stand and permit mounting with intromission.

In the spider monkey *(Ateles geffroyi* and *Ateles belzebuth),* visual communication is an important element of the interaction pattern. Vocalizations are important, but do not appear to play the same role as in a shrew encounter. Direct approach with subsequent avoidance characterizes an initial encounter between animals unfamiliar to one another. The tabulation of responses differs from the shrew since the direct approach and moving away are approximately reciprocal responses; e.g., the spider monkey does not broadcast his nearness by sound emission, but rather through the act of moving toward a partner. Although chemical communication in spider monkeys, involving sternal glands [EISENBERG and KUEHN, 1966; EPPLE and LORENZ, 1967] and urine [KLEIN, 1971], is of some significance, the role of chemical communication in other primate species remains to be clarified.

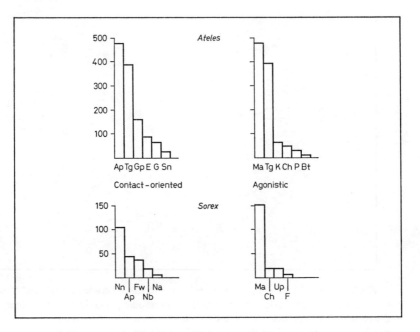

Fig. 2. Comparison in monkey and shrew of frequency distributions for behavior patterns and configurations commonly shown in two-animal encounters. Ap = Approach; Tg = touch or tag; Gp = grapple; E = embrace; G = grasp; Sn = sniff anogenital area; K = kick; Ch = chase; P = push; Bt = bite; Nn = nose to nose; Fw = follow; Nb = nose to body; Na = nose to anogenital area; Up = upright with warding; F = flight. (Sounds and expressions omitted.) The distributions clearly show a similarity between monkey *(Ateles)* and shrew *(Sorex)* in the number of elements commonly exhibited during a two-animal interaction. The interaction patterns are broken, somewhat arbitrarily, into two subclasses: contact-oriented and agonistic (see text for explanation). Data based on 29 two-animal encounters averaging 30 minutes each for *Ateles* and 30 two-animal encounters of 3-minutes duration for *Sorex.*

The encounter behavior of two spider monkeys will include much initial approaching and moving away and will finally result in establishment of contact, generally involving reaching out and touching, which may lead to an embrace and mutual sniffing of pectoral glands or may develop into agonistic behavior with rather aggressive slapping and kicking, leading, in turn, perhaps, to an actual chase with flight on the part of the submissive animal. On the other hand, embracing and pectoral sniffing may lead to a form of play-fighting, termed grappling, that involves an introductory head-shaking movement followed by a complicated series of wrestling movements often accompanied by a characteristic vocalization [EISENBERG and KUEHN, 1966].

Although the encounter behavior of shrews and monkeys under these circumstances differs qualitatively, it involves a similar number of behavior elements (fig. 2). Furthermore, the frequency of the behavioral elements seems to follow a general rule. Frequency of actual contact is lower than non-specific responses, including approach or moving away. Actions that involve expenditure of a great deal of energy, such as fighting in the case of shrews, biting or grappling in the case of spider monkeys, or intimate contact exchanges with naso-anal-genital contact are of a very rare occurrence. It would appear that the distribution of the acts in terms of their frequency somewhat conforms to ZIPF's [1949] 'law of least effort' since the most nonspecific actions requiring the least expenditure of muscular energy occur with the greatest frequency.

In presenting data in figure 2, I have omitted tabulations of facial expressions and vocalizations in the spider monkey and vocalizations in the insectivore. Certain movement patterns of the shrew have potential visual communicatory significance; but considering the reduced eye of the shrew, one can only conclude that these postures do not convey any visual information. Experimental evidence with respect to this point is included in the discussion by GOULD [1969]. Furthermore, my tabulation involves macro-elements of behavior. For example, one could take the behavior pattern of grappling in the spider monkey and break it down into its components, such as reaching out, touching, holding the fur, and mock biting, and could analyze the individual foot and arm movements. I must say, however, that most behavioral inventories are not that 'fine grained'. And, if one were to do this fine-grain analysis on grappling in spider monkeys, one would be forced also to do an equally fine-grain analysis on the typical rolling fight or upright and warding reactions of shrews that would probably result in the maintenance of an equivalent number of motor units in the repertoires of the two species. Of course, the spider monkey has the potential of transmitting a great deal of information via the visual channel of communication, a mode that leads to the next section.

The Complexity of Communication Systems

The previous discussion leads inevitably to a consideration of information exchange and those motor elements and patterns in a mammalian interaction that are involved in the transfer of information: the movements, deposition of chemical traces, utterance of vocalizations, body postures, and facial expressions. As a word of caution, I should say that the description of stereo-typed movement patterns and expressions with presumptive communicatory value has progressed much further with birds and fishes than with primates. These taxa are governed in the main by visual input, with olfaction as a secondary possibility in fishes but virtually absent in most birds. On the other hand, many present-day mammals, as well as the stem forms in the evolutionary sense, were probably nocturnal or crespuscular; hence, there is an ancient mammalian heritage where olfaction, audition, and tactile modes of communication are pre-eminent. Refined visual discrimination was probably unimportant in the early mammalian line of evolution. Many recently evolved taxa of mammals, including the higher primates, ungulates, and higher carnivores, have refined their visual capacity; and visual communication modes are here more dominant than in the more generalized or primitive forms. However, the ancient heritage of olfaction and audition, plus that all-important tactile input, still remains as part and parcel of the mammalian interaction system.

We have unfortunately concentrated very much in the early studies of mammalian ethology on attempts to define displays that are dependent on the visual mode of input but may not be all that important in the coordination of mammalian interactional systems. Perhaps the concept of display in the classical sense may not be too applicable to a mammalian condition. In an interactional analysis of mammals, it is extraordinarily important to note the orientation of the major sense organs of both the presumptive sender and the receiver. In addition, one has to note the general conformation and alignment of the body, the orientation of the limbs and tail, and the movement patterns of the two interacting animals with respect not only to one another but also to artifacts in the environment [GOLANI and MENDELSSOHN, 1971; WEMMER, 1972]. Indeed, alignment of body axes and general conformation of the body may be of extreme importance in forms that communicate visually; but since, except for two or three cases, we lack adequate descriptions to permit an analysis that takes account of configurational and gait-associated patterns, I fear we must still rely on our rather incomplete behavioral inventories.

Let us look at several non primate species to gain some perspective on the communication problem. The Asiatic elephant *(Elephas maximus)* produces three to four fundamental sound types, and, within each type, there are two to three variations yielding ten readily recognized vocalization patterns [McKAY, 1973]. A basic set of ten fundamental vocalization patterns compares well with *Gorilla gorilla* [SCHALLER, 1963]. The elephant can indicate mood change by variations in the position of the trunk and ears involving a graded series of expressions, and some twenty trunk and ear combinatorials can be discerned [KÜHME, 1962]. Thus, elephants have the possibility of combining at least ten basic sound types with twenty trunk-ear configurations; and this total discounts body orientation and overall gait-associated movement. I think we would agree that this potential communicatory significance compares favorably with most primates, in spite of the fact that the elephant is almost unable to change its facial expression as such.

In turning to nonprimate species that have the potential for facial expressions in a graded series, let us consider two carnivores, the semi-solitary domestic cat *(Felis catus)* and the rather social wolf *(Canis lupus)*. The cat produces approximately eight distinct vocalizations with some intermediates; there are three fundamental forms of chemical marking, yielding at least three different chemical signals. And there are at least nine facial expressions forming a graded series that may be combined with 16 tail and body postures, giving a total of at least 16 elements in combination [TEMBROCK, 1970; LEYHAUSEN, 1956]. The wolf *(Canis lupus)* also has at least eight distinguishable vocalizations, three forms of chemical marking, at least ten distinct facial expressions, and at least 16 tail and body postures that may be combined to give a total of at least 16 configurations in a graded series [M. W. FOX, 1971; KLEIMAN, 1966; SCHENKEL, 1948]. Thus, so-called solitary carnivores do not differ radically from social carnivores with respect to their effective repertoires.

Some may argue that recognizable elements within graded series, although of some communicatory significance, raise problems concerning the definition of what an element is; thus, we must consider the problem of a display, for which I turn to MOYNIHAN [1967, 1970]. For MOYNIHAN, the term display is restricted to a movement pattern of an animal or a vocalization that conveys not only information but also exhibits some ritualization or stereotypy; that is, the movement pattern has become specialized in form and frequency as an adaptation to permit or facilitate communication. Very often a display has several components within it; one component may be relatively stereotyped, although it should not be thought that all displays conform to the criterion of a fixed action pattern [see BARLOW, 1968]. If we accept MOYNIHAN's definition of display and consider his theoretical paper of 1970, we find that, by his own tabulation, the displays of primates do not differ significantly from the displays of other ordinal groups of mammals. For example, according to MOYNIHAN, the average number of displays for some 24 mammalian species, that have been studied in detail, is 21.54 displays per species. For eight species of primates studied, the range is 16 to 37 with an average of 27.25. For five species of *Artiodactyla*, the range is 13 to 26 with an average of 22. Further, the average for 26 species of birds is 20.38 displays, and for 12 species of fish, 16.16. One primate species in Moynihan's list seems to have the greatest number of displays recorded, only because the particular study involved a rather fine-grained analysis of vocal signals. Using Moyni-han's method of identifying displays, we certainly do not find any startling departures on the part of primates from the higher vertebrate norm.

If I redefine the potential communicatory mechanisms into the categories of vocalizations, facial expressions and gestures, chemical marking, move-ments and configurations, and orientation movements, then I can establish the following comparisons between four genera of shrews *(Blarina, Suncus, Cryptotis,* and *Sorex)* on the one hand [Gould, 1969; Eisenberg, 1964], and one species of primate *(Ateles geoffroyi)* on the other.

For *Ateles,* one can define five fundamental vocalization-moods, within which there are one to five subvariants. A possible high count of 17 vocaliz-ation forms can be identified, of which 13 appear to be rather stereotyped in their form. Major facial expressions and gestures include six that may be combined with vocalization to give 30 observed combinations, ten of which, however, predominate in social interactions. Chemical marking involves two predominant modes, as discussed previously. Movements and relatively stereotyped interaction configurations add up to about ten, if we omit the loco-motory variants. Orientation movements to a partner involve five discernible types.

For the shrew species, vocalizations involve more than 12 variations based on six fundamental moods [for an expanded discussion, see Gould, 1971]. Facial expressions are virtually absent, so do not enter into combination with vocalizations; in any event, these animals have very reduced visual acuity. The movements and configurations add up to nine discernible units that may, however, be of no visual significance. Chemical marking involves at least three methods of scent deposition. Orientation configurations with respect to a partner total five.

Thus there is only one aspect of potential communication in which *Ateles* shows a more refined set of signals than the shrews. *Ateles* is able to combine six major facial expressions and head gestures with 13 elemental types of vocalizations; ten of these combinations occur frequently enough to be as-signed a contextual communication value. What is surprising for us at this point is that the vocal repertoire of the shrew appears to be potentially at a level of complexity commensurate with that of the spider monkey. It remains to be demonstrated, however, that complex information is actually trans-mitted. Given the spider monkey's potential for visual communication then, we can certainly state that it has the capacity to communicate information via

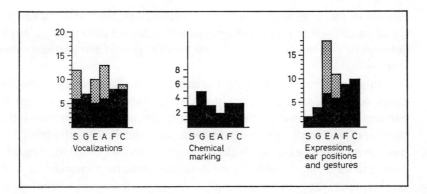

Fig. 3. Comparisons of communication repertoires for selected species of mammals. E = *Elephas maximus;* G = *Genetta tigrina;* A = *Ateles geoffroyi;* S = soricoid insectivores; F = *Felis catus;* C = *Canis lupus.* Black = fundamental patterns; stippled = discernible graded elements, if noted.

this channel that the shrew does not. We should take scant comfort in this, however, since the spider monkey's potential methods of transmitting information visually are no more complex than those already outlined for the elephant and some carnivores (see fig. 3).

The Formation of Tradition

It has been widely assumed that higher primates exhibit learned traditions and that this characteristic sets them apart from the behavior patterns of other mammalian species. I should point out that the learning of traditions with respect to song development in passeriform birds is an active area of research [HINDE, 1969]. Dialect formation in birds is an established fact, and there is no reason to assume that dialect formation is the prerogative of a big-brained mammal. There is no reason why dialect formation cannot or has not occurred in a wide variety of mammalian species, but for mammals it is a relatively unresearched area [MARLER and TAMURA, 1964; LeBOEUF and PETERSON, 1969].

The donation of a constructed artifact to subsequent generations is a well-known phenomenon that was given its most rigorous analysis with respect to successive burrow occupancy by generations of the common Norway rat *(Rattus norvegicus)* [CALHOUN, 1963]. It is difficult to appreciate the amount of energy expended in burrow construction, but CALHOUN has prepared an analysis that convincingly demonstrates the advantage gained by subsequent generations of rats if they 'inherit' part of the parental tunnel systems.

In a conservative mammalian species such as *Bradypus infuscatus*, the three-toed sloth, we find not only traditional learning of home range use patterns but donation of part of a home range on the part of a mother sloth at the time of weaning her offspring. MONTGOMERY and SUNQUIST [in press] have noted in an extensive field study that each young sloth is carried by the mother for five to six months after birth and during this time the young climbs on the mother, feeds, but continues to maintain physical contact. Although the young appear to be nutritionally weaned at about a month of age, they do not leave the mother until some five months later. The final departure from the mother is termed 'social weaning'; at this time the mother disengages from the young and moves away from it to a new part of her home range. The young continues to move in a part of the home range that it had become familiar with in the course of its travels with the mother; thus, the young appears to inherit a portion of the mother's home range and has learned the successive use of tree species based on feeding preferences established while accompanying the mother in her own feeding pattern.

There is ample documentation that ungulates, such as the mountain sheep *(Ovis canadensis* and *Ovis dalli)*, have learned traditions of seasonal habitat utilization patterns [GEIST, 1971]. Similar phenomena can be demonstrated for *Elephas maximus* [MCKAY, 1973]. The list could go on and on; but, in summary, it would appear that any reasonably large, long-lived mammal exhibiting considerable generation overlap in its social group composition has approximately an equivalent capacity for the formation of tradition.

Complexity of Social Organization

It has been assumed that higher primate social organizations are infinitely more complex than those of lower mammals. Figure 4 compares the composition of a communal group of tenrecs *(Hemicentetes semispinosus)* with a matriarchal herd of *Loxodonta africana*, the African elephant. When the time scale of the tenrec is adjusted to correspond to the long-lived elephant, one can plainly see that there are stages of social complexity within its communal burrows that are equivalent to those shown by the elephant. Only the time scale differs. By any objective measure of complexity, once time scales

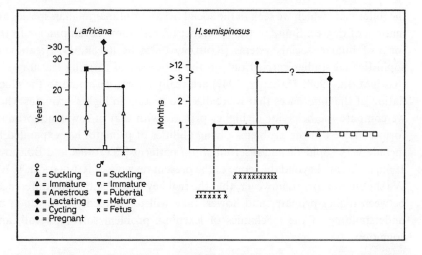

Fig. 4. Composition of a family group of elephants *(Loxodonta africana)* as portrayed in LAWS and PARKER [1968], compared to a similar configuration shown by the streaked tenrec *(Hemicentetes semispinosus)* [EISENBERG and GOULD, 1971].

have been adjusted for life cycle differences, the social structures of higher primates are no more complex than those portrayed by many mammals [EISENBERG and GOULD, 1970].

Those differences observed in the behavior of higher primates which appear to set them off distinctly from most other mammals are not the prerogatives of higher primates alone [EISENBERG and KUEHN, 1966]. The differences result from the fact that higher primates and other large, long-lived mammalian species (e. g., elephants, cetaceans, and large carnivores) are able to store and retrieve a great amount of information collected on an individual basis over a long period of time. Hence, greater individual response forms to social and physical situations can be discerned in such species, which have a greater capacity for individual recognition and for the formation of individualistic social relationships [KUMMER, GARTLAN, this volume].

KUMMER [1968] has amply demonstrated for *Papio hamadryas* that individual recognition and individualistic response levels are demonstrable in baboon society, giving rise to variations on a common theme in their social structure based on a one-male group. MENZEL [1971] has shown the rudiments of symbolic communication present in the chimpanzee *(Pan satyrus)* that allow it to cope with complicated detour problems and allow animals to impart information about the state of an environment to partners. However, the differences which we seek in the social behavior of mammalian species are matters of degree. Some of the differences foreshadow developments in the form of human social systems [KÖHLER, 1925; SCHILLER, 1952]; and the sophisticated studies carried out on the processes of forming learning sets [RUMBAUGH, 1968; HARLOW, 1949] are really efforts germaine to the elucidation of the differences that we realize are present in higher primates when we compare the behavior of higher primate with that of lower nonprimate forms. Interest in the problem-solving abilities of primates has expanded to include very sophisticated discrimination patterns [FLETCHER and BORDOW, 1965] which are beyond the scope of the present review [see RUMBAUGH, 1970]. We should not forget, however, that the real key to understanding differences between higher primates and 'lower' taxa will undoubtedly pivot upon an understanding of the mechanics of learning, problem solving, recall, and cognition.

Summary

This paper is of necessity an overview and rather simplified. I have tried to demonstrate that some preconceived notions about the nature of primate sociality are without validity. This effort does not mean that higher primate behavior does not differ in degree or quality from that of other mammalian species; only that we have been seeking to prove things or have taken for granted assumptions that are unwarranted. There are real and genuine differences between the behavior of a troop of *Ateles geoffroyi* and an interacting pair of shrews, but the reasons that we see differences or suspect differences should not be based upon superficial notions of complexity of repertoire and complexity of interaction; rather, they are related to our recognition of the ability of primates to assemble information acquired on an individualistic basis and in turn to apply it to specific and unique situations. I am not suggesting that we abandon primate studies in favor of studies of rodents and insectivores. What I am encouraging is a broad base of research on animals (and not just mammals), coupled with a broad frame of comparisons. Comparisons between insects and vertebrates can provide fruitful contrasts and insights of as yet unknown heuristic value [WILSON, 1971].

Staying within the class *Mammalia*, I could not hope for more than to encourage those new students of behavior interested in field studies to concentrate on more nonprimate species that are large and long-lived and have rather large brains and an increased capacity for learning. To this end one can profitably study elephants, dolphins, ungulates, and large carnivores. In particular, I would like to encourage new workers to consider the value of carrying on long-term studies with a large carnivore such as the bear. Bears are almost unexplored in terms of their learning capacities and in terms of their sociology[2]. Here in North America we have in *Ursus middendorfi* the largest terrestrial carnivore, a form that remains almost totally neglected as an object of study. One should not be put off because the bear is semi-solitary, since such species have rich repertoires for communication and possess a poorly understood yet viable communal life [LEYHAUSEN, 1965]. Once again let me stress that new insights are always forthcoming by comparisons and that comparison in the last analysis is one of the best and only tools known to generate such insights.

References

ALTMANN, S.A.: A field study of the sociobiology of rhesus monkeys, *Macaca mulatta*. Ann. N.Y. Acad. Sci. *102:* 338–435 (1962).

BARLOW, G.: Ethological units of behavior; in INGLE The central nervous system and fish behavior, pp. 217–232 (University of Chicago Press, Chicago 1968).

CALHOUN, J.B.: The ecology and sociology of the Norway rat. US publ. Hlth Serv. Publ. No. 1008 (US Government Printing Office, Washington 1963).

CHOMSKY, N.: Aspects of the theory of syntax (MIT Press, Cambridge 1965).

DEVORE, I.: The evolution of human society; in EISENBERG and DILLON Man and beast: comparative social behavior. Smithsonian Annual III, pp. 299–311 (Smithsonian Institution, Washington 1971).

EIBL-EIBESFELDT, I.: Das Verhalten der Nagetiere; in Handbuch der Zoologie, vol. VIII/10, 13, pp. 1–88 (1958).

EISENBERG, J.F.: Studies on the behavior of *Sorex vagrans*. Amer. Midl. Natur. *72:* 417–425 (1964).

EISENBERG, J.F.: Comparative studies on the behavior of rodents with special emphasis on the evolution of social behavior. Part I. Proc. US nat. Mus. *122 (3597):* 1–55 (1967).

EISENBERG, J.F.: Behavior patterns; in KING Biology of *Peromyscus* (Rodentia). Amer. Soc. Mammal., special publ. *2:* 451–459 (1968).

EISENBERG, J.F.: Social organization and emotional behavior; in TOBACH Experimental approaches to the study of emotional behavior. Ann. N.Y. Acad. Sci. *159:* 752–760 (1969).

EISENBERG, J.F. and GOULD, E.: The tenrecs: a study in mammalian behavior and evolution. Smithsonian Contrib. Zool. *27:* 1–137 (1970).

2   A current summary and reference work is: *Bears – Their Biology and Management.* (1972) I.U.C.N. Publ. No. 23, 371 pp.

EISENBERG, J.F. and KUEHN, R.E. The behavior of *Ateles geoffroyi* and related species. Smithsonian misc. Coll., vol. 151 (8), publ. No. 4683 (Smithsonian Institution, Washington 1966).

EPPLE, G. und LORENZ, R.: Vorkommen, Morphologie und Funktion der Sternaldrüse bei den Platyrrhini. Folia primat. *7:* 98–126 (1967).

FLETCHER, H.J. and BORDOW, A.M.: Monkey's solution of an ambiguous-cue problem. Percept. Motor Skills *21:* 115–119 (1965).

FOX, M.W.: Behaviour of wolves, dogs and related canids (Jonathan Cape, London 1971).

FOX, R.: The cultural animal; in EISENBERG and DILLON Man and beast: comparative social behavior. Smithsonian Annual III, pp. 273–297 (Smithsonian Institution, Washington 1971).

GARDNER, R.A. and GARDNER, B.T.: Teaching sign language to a chimpanzee. Science *165:* 664–672 (1969).

GEIST, V.: Mountain sheep: A study in behavior and evolution (University of Chicago Press, Chicago 1971).

GOLANI, I. and MENDELSSOHN, H.: Sequences of precopulatory behavior of the jackal *(Canis aureus L.)* Behaviour *38:* 169–192 (1971).

GOULD, E.: Communication in three genera of shrews *(Soricidae): Suncus, Blarina,* and *Cryptotis.* Comm. behav. Biol. A *3:* 11–31 (1969).

GOULD, E.: Studies of maternal-infant communication and development of vocalizations in the bats *Myotis* and *Eptesicus.* Comm. behav. Biol. *5:* 263–313 (1971).

HARLOW, H.F.: The formation of learning sets. Psychol. Rev. *56:* 51–65 (1949).

HINDE, R.A. (ed.): Bird vocalizations (Cambridge University Press, Cambridge 1969).

JAYNES, J. and BRESLER, M.: Evolutionary universals, continuities and alternatives; in EISENBERG and DILLON Man and beast: comparative social behavior. Smithsonian Annual III, pp. 333–346 (Smithsonian Institution, Washington 1971).

KLEIMAN, D.: Scent marking in the *Canidae.* Symp. zool. Soc., Lond. *18:* 167–177 (1966).

KLEIN, L.L.: Observations on copulation and seasonal reproduction of two species of spider monkeys, *Ateles belzebuth* and *A. geoffroyi.* Folia primat. *15:* 233–248 (1971).

KÖHLER, W.: The mentality of apes (Harcourt & Brace, New York 1925).

KÜHME, W.: Ethology of the African elephant (*Loxodonta africana* Blumenbach 1797) in captivity. Int. Zoo Yb. *iv:* 113–121 (1962).

KUMMER, H.: Social organization of Hamadryas baboons: a field study (University of Chicago Press, Chicago 1968).

LAWICK-GOODALL, J. VAN: Tool-using in primates and other vertebrates; in LEHRMAN, HINDE and SHAW Advances in the study of behavior, pp. 195–249 (Academic Press, New York 1970).

LAWS, R.M. and PARKER, I.S.C.: Recent studies on elephant populations in East Africa, Symp. zool. soc., Lond. *21:* 319–359 (1968).

LEBOEUF, B.J. and PETERSON, R.S.: Dialects in elephant seals. Science *166:* 1654–1656 (1969).

LEYHAUSEN, P.: Verhaltensstudien an Katzen. Z. Tierpsychol. Beiheft *2:* 1–120 (1956).

LEYHAUSEN, P.: The communal organization of solitary mammals. Symp. zool. Soc., Lond. *14:* 249–263 (1965).

LORENZ, K.: Part and parcel in animal and human societies; in LORENZ Studies in animal and human behavior, vol. 2, pp. 115–195 (Harvard University Press, Cambridge 1971).

MARLER, P.: Familiarity and strangeness as determinants of animal allegiance and hostility; in RICHARDSON Mechanisms of allegiance and hostility (University of Virginia Press, Charlottesville, in press).

MARLER, P. and TAMURA, M.: Culturally transmitted patterns of vocal behavior in sparrows. Science 146: 1483–1486 (1964).

McKAY, G. M.: Behavior and ecology of the elephant (Elephas maximus) in southeastern Ceylon. Smithsonian Contrib. Zool. No. 125 (1973).

MENZEL, E. W., jr.: Communication about the environment in a group of young chimpanzees. Folia primat. 15: 220–232 (1971).

MENZEL, E. W., jr.: Spontaneous invention of ladders in a group of young chimpanzees. Folia primat. 17: 87–106 (1972).

MONTGOMERY, G. G. and SUNQUIST, M. E.: Social weaning, home range inheritance, and home ranges of three-toed sloths. Smithsonian Contrib. Zool. (in press).

MORRIS, D.: The biology of art (Methuen, London 1962).

MOYNIHAN, M.: Comparative aspects of communication in New World primates; in MORRIS Primate ethology, pp. 236–266 (Aldine, Chicago 1967).

MOYNIHAN, M.: Control, suppression, decay, disappearance and replacement of displays. J. theor. Biol. 29: 85–112 (1970).

PAGET, R.: Human Speech (Paul French, Truber & Co., London 1930).

RENSCH, B.: Die Wirksamkeit ästhetischer Faktoren bei Wirbeltieren. Z. Tierpsychol. 15: 447–461 (1958).

RUMBAUGH, D. M.: The learning and sensory skills of the squirrel monkey in phylogenetic perspective; in ROSENBLUM and COOPER The squirrel monkey, pp. 255–317 (Academic Press, New York 1968).

RUMBAUGH, D. M.: The learning skills of anthropoids; in ROSENBLUM Primate behavior, vol. 1, pp. 1–70 (Academic Press, New York 1970).

SCHALLER, G. B.: The mountain gorilla (University of Chicago Press, Chicago 1963).

SCHENKEL, R.: Ausdruck-Studien an Wölfen. Behaviour 1: 81–129 (1948).

SCHILLER, P. H.: Innate constituents of complex responses in primates. Psychol. Rev. 59: 177–191 (1952).

TEMBROCK, G.: Bioakustische Untersuchungen an Säugetieren des Berliner Tierparks. Milu 3: 78–96 (1970).

TITIEV, M.: The Science of man (Holt, New York 1955).

WEMMER, C.: Comparative ethology of the large-spotted genet, Genetta tigrina, and related viverrid genera; Ph.D. thesis, University of Maryland, College Park (1972).

WILSON, E. O.: The insect societies (The Belknap Press, Harvard University Press, Cambridge 1971).

ZIPF, G. K.: Human behavior and the principle of least effort (Addison-Wesley Press, Cambridge 1949).

Author's address: Dr. JOHN EISENBERG, Resident Scientist, Department of Zoological Programs, National Zoological Park, Smithsonian Institution, Washington, DC 2009 (USA)

Authors' Index

Subject Index